全国高等院校云计算系列"十四五"规划教材

Docker
容器技术实战项目化教程

主　编◎杨建清　陈小明　柏杏丽
副主编◎朱家荣　韩　冰　殷　瑛　李春平　张　锴
参　编◎王俊波　刘红兵　马玉芳　王彩峰

中国铁道出版社有限公司
CHINA RAILWAY PUBLISHING HOUSE CO., LTD.

内 容 简 介

本书基于 Ubuntu 18.04 环境搭建企业场景，将真实场景与配置实例紧密结合，使读者能够快捷、直观、深刻地掌握 Docker 容器技术的相关知识和操作技能，增强实战经验。

本书主要内容包含两方面，一方面是单主机模式下的 Docker 容器技术实现，包含项目一的认识 Docker，项目二的学习 Docker 镜像，项目三的管理容器外加数据卷，项目四的认识和理解 Docker 网络，项目五的使用 Compose 编排服务；另一方面是多主机模式，搭建真实生产服务环境，由 3 台主机实现，包括项目六的认识和理解多主机网络，项目七的使用 docker-swarm 编排网络服务和项目八的使用 Kubernetes 编排网络服务。

本书从零基础开始，逐步深入学习容器技术，适合作为普通高等院校本科计算机网络、云计算等相关专业的容器技术相关课程的教材，也可作为 Docker 容器基础和容器编排技术的自学参考书。

图书在版编目（CIP）数据

Docker 容器技术实战项目化教程 / 杨建清，陈小明，柏杏丽主编 . —北京：中国铁道出版社有限公司，2021.3（2025.1 重印）
全国高等院校云计算系列"十四五"规划教材
ISBN 978-7-113-27629-4

Ⅰ. ①D… Ⅱ. ①杨…②陈…③柏… Ⅲ. ①Linux 操作系统－程序设计－高等学校－教材 Ⅳ. ①TP316.85

中国版本图书馆 CIP 数据核字（2020）第 273164 号

书　　名：Docker 容器技术实战项目化教程
作　　者：杨建清　陈小明　柏杏丽

策　　划：韩从付　　　　　　　　　　　　　　编辑部电话：（010）63549501
责任编辑：贾　星　包　宁
封面设计：刘　莎
责任校对：孙　玫
责任印制：赵星辰

出版发行：中国铁道出版社有限公司（100054，北京市西城区右安门西街 8 号）
网　　址：https://www.tdpress.com/51eds
印　　刷：北京市泰锐印刷有限责任公司
版　　次：2021 年 3 月第 1 版　2025 年 1 月第 4 次印刷
开　　本：787 mm×1 092 mm　1/16　印张：16.25　字数：426 千
书　　号：ISBN 978-7-113-27629-4
定　　价：49.80 元

版权所有　侵权必究

凡购买铁道版图书，如有印制质量问题，请与本社教材图书营销部联系调换。电话：（010）63550836
打击盗版举报电话：（010）63549461

前言

当前信息技术正从传统IT架构迁移至云原生架构，从虚拟化到云计算，从虚拟机到容器，从微服务到无服务器计算，技术的迭代前所未有，计算机网络运维和应用程序开发也发生了翻天覆地的变化，其中微服务架构中的容器技术成为云原生的重要组成部分。

在容器技术中，Docker技术成为当前热门的技术。Docker让开发工程师可以将其应用和依赖封装到一个可移植的容器中，而不用关心底层操作系统。Docker几乎可以解决虚拟机能够解决的所有问题，还能够解决虚拟机由于资源要求过高而无法解决的问题。

服务编排是云原生架构的重要特征，通过集中式的编排调度系统来动态地管理和调度服务。Kubernetes、Mesos和Docker Swarm都是典型的编排系统，其中Kubernetes脱颖而出，可以实现让容器应用进入大规模工业生产。

本书以Ubuntu 18.04操作系统为基础，尽量做到从零基础开始，逐步深入学习容器技术，因此内容的编排从安装Linux操作系统开始，然后学习安装Docker、Docker镜像技术、数据卷挂载、Docker网络、Docker的Compose编排技术、多主机环境搭建、多主机环境下的Docker网络搭建、Swarm编排技术，以及Kubernetes编排技术。

本书由杨建清、陈小明、柏杏丽任主编，由朱家荣、韩冰、殷瑛、李春平、张锴任副主编，王俊波、刘红兵、马玉芳、王彩峰参与编写。

本书是广东白云学院、广州市白云工商技师学院与云宏信息科技股份有限公司校企合作、资源共建的成果之一。在编写过程中，云宏信息科技股份有限公司提供了大量的实际生产案例，技术工程师提供了技术解决方案，在此一并表示感谢。

本书概念部分参考了互联网上公布的一些资料。由于互联网上资料较多，引用复杂，无法一一注明原出处，原文版权属于原作者所有。其他参考文献在本书后列出。

由于编者水平有限，书中难免存在疏漏和不妥之处，恳请读者批评指正，以期修订时进一步完善。

编　者

2020 年 12 月

目 录

项目一 认识 Docker 1
    任务一   安装和配置操作系统 1
    任务二   运行第一个测试容器 20
    任务三   掌握容器的基本操作 25

项目二 学习 Docker 镜像 30
    任务一   认识 Docker 镜像分层 30
    任务二   利用 dockerfile 生成镜像 37
    任务三   搭建本地镜像仓库 40
    任务四   创建加密的私有仓库 45
    任务五   使用官方公共镜像 Registry 50
    任务六   对镜像和容器进行打包 53

项目三 管理容器外加数据卷 62
    任务一   通过宿主机目录挂载容器数据卷 62
    任务二   通过卷容器挂载数据卷 74

项目四 认识和理解 Docker 网络 82
    任务一   认识 Docker 网络 82
    任务二   分析自定义 Docker 网络 90

项目五 使用 Compose 编排服务 105
    任务一   使用 Compose 实现高可用 Web 网站建设 105
    任务二   使用 Compose 实现个人博客网站建设 110
    任务三   运用 Compose 使用现有镜像配置 LNMP 网站 116
    任务四   使用 Compose 编译实现 LNMP 网站建设 122

项目六 认识和理解多主机网络 134
    任务一   安装配置 docker-machine 134
    任务二   分析多主机 overlay 网络 141

项目七	使用 docker-swarm 编排网络服务	168
	任务一　认识 docker-swarm	168
	任务二　利用 docker-swarm 创建 Nginx 集群	177
	任务三　运用 Docker Stack 部署 LNMP 网站	194
	任务四　运用 Docker Stack 部署高可用个人博客网站	208
项目八	使用 Kubernetes 编排网络服务	215
	任务一　认识 Kubernetes 架构	215
	任务二　部署和测试 Kubernetes 集群	225
	任务三　通过 NFS 网络卷部署 Kubernetes Nginx 集群服务	238
	任务四　通过 PV 和 PVC 部署 Kubernetes Nginx 集群服务	244

项目一 认识 Docker

【项目综述】

根据公司的业务需求，公司的业务将移植到微服务架构的容器上，需要选取软硬件环境，安装和配置 Docker 环境，测试并成功运行一个容器。

【项目目标】

◎ 能够配置容器（Docker）运行环境

◎ 能够测试运行一个 Docker 容器（http）

任务一　安装和配置操作系统

任务场景

由于业务需求，安安公司想将自己信息系统的业务移植到容器（Docker）上面。现在技术人员需要创建一个测试环境，随后逐步将业务移植到容器（Docker）上面。假设你是云计算工程师，需要你打造该环境。经过公司信息部的研讨，决定采用Ubuntu 18.04操作系统作为底层OS，然后逐步安装完善。

任务描述

本任务学习安装Ubuntu 18.04操作系统，采用VMware Workstation虚拟机模拟实现，网卡采用NAT地址转换，由于业务需要，内存设置为16 GB，CPU为4颗，硬盘为120 GB，要求安装好系统并设置好IP地址，能与互联网通信，设置apt源，能实现系统更新。

任务目标

◎ 能够在VMware Workstation环境中创建虚拟机

◎ 能够根据需求设置虚拟机配置参数

◎ 能够安装Ubuntu 18.04操作系统，登录测试成功

◎ 能够配置网络并测试网络成功

任务实施

一、用 VMware 创建虚拟机

1. 新建虚拟机

打开"新建虚拟机向导"对话框，选择安装方式，如图1-1所示。这里选择"典型"（推荐）安装方式。单击"下一步"按钮，进入选择安装来源界面。

2. 选择光盘镜像

选择准备好的Ubuntu 18.04光盘镜像，如图1-2所示。单击"下一步"按钮，进入设置用户名界面。

图1-1 选择虚拟机安装方式

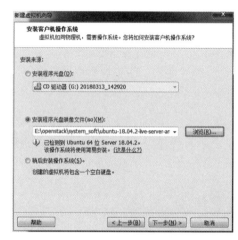

图1-2 选择光盘镜像

3. 设置用户名

设置安装系统的用户名（根据自己的规划设置），如图1-3所示。单击"下一步"按钮，进入设置虚拟机名称界面。

4. 设置虚拟机的名称

根据自己的规划设置主机的名称和存放位置，注意安装的磁盘需要有足够空间，如图1-4所示。单击"下一步"按钮，进入设置磁盘空间界面。

图1-3 设置用户名

图1-4 设置虚拟机的名称和存放位置

5. 设置磁盘空间

设置磁盘大小和存储文件方式，这里磁盘空间设为120 GB，随后预装Docker相关服务，如图1-5所示。当然也可以小一点，但是建议设置磁盘空间不要小于80 GB。单击"下一步"按钮，进入设置内存大小界面。

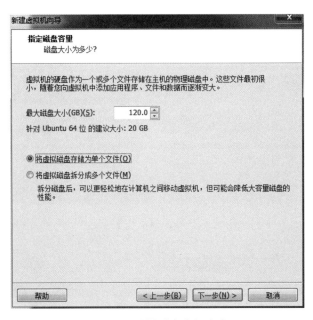

图1-5　设置磁盘空间大小

6. 设置内存大小

根据需求设置内存大小，内存应尽量大，如图1-6所示。建议不要小于8 GB。

图1-6　设置内存大小

7. 设置处理器

设置CPU数量，按照宿主机资源情况设置，此处可以将虚拟化打开。本任务Docker是不需要打开虚拟化的，但为了方便以后安装其他虚拟化，此处打开虚拟化设置，选中"虚拟化Intel VT-x/EPT或AMD-V/RVI"复选框，如图1-7所示。单击"关闭"按钮。

图 1-7　设置处理器

8. 设置完成

设置完成，列出配置清单，如图1-8所示。单击"完成"按钮，至此虚拟机的设置完成，接下来就可以安装Ubuntu 18.04系统了。

图 1-8　虚拟机设置完成界面

9. 完成虚拟机创建

单击"完成"按钮，可以看到创建的虚拟机，如图1-9所示，可以安装Ubuntu 18.04系统。

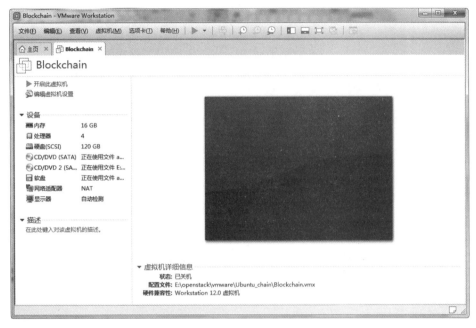

图1-9　完成虚拟机创建

二、安装 Ubuntu 18.04 操作系统

这里选择英文方式安装，为了方便读者阅读，界面上出现的英文在后面的括号中翻译为中文。

1. 选择语言

这里选择English，如图1-10所示。

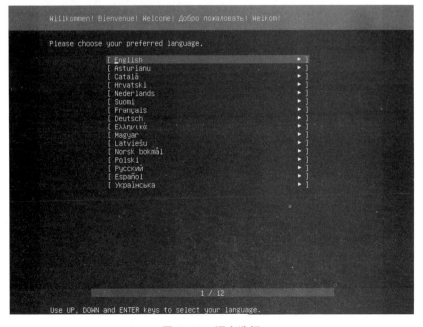

图1-10　语言选择

2. 键盘配置

键盘配置界面如图1-11所示。

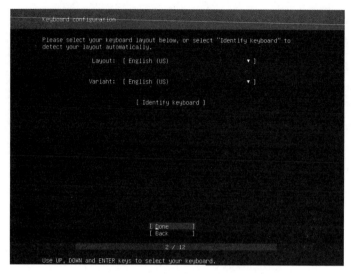

图1-11 键盘配置

Keyboard configuration（键盘配置）

Please select your keyboard layout below,or select "Identify keyboard" to detect your layout automatically.（请在下面选择键盘布局，或选择"识别键盘"自动检测布局。）

这里保持默认设置，直接按【Enter】键。

3. 安装系统

如图1-12所示，界面出现欢迎语：

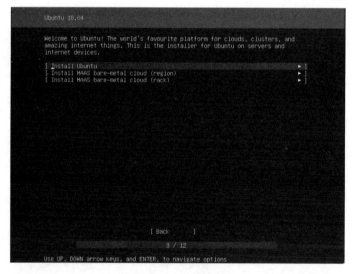

图1-12 安装菜单

Welcome to Ubuntu! The world's favourite platform for clouds, clusters, and amazing internet things. This is the installer for Ubuntu on servers and internet devices.（欢迎来到Ubuntu！世界上最受欢迎的云平台、集群和惊人的互联网事物。这是服务器和Internet设备上Ubuntu的安装程序。）

下面有3个安装选项：

Install Ubuntu

Install MAAS bare-metal cloud (region)

Install MAAS bare-metal cloud (rack)

这里选择Install Ubuntu选项，另外也可以通过MAAS技术安装。

4. 网络连接配置

配置网络连接界面如图1-13所示。下面是对界面的解释。

Network connections（网络连接）

Configure at least one interface this server can use to talk to other machines, and which preferably provides sufficient access for updates.（配置此服务器可用于与其他计算机对话的至少一个接口，该接口最好为更新提供足够的访问权限。）

这里默认通过dhcp方式获取到一个IP地址192.168.47.132，直接选择Done执行下一步。也可以选择ens32网卡，然后选择手动方式设置网段、IP地址、子网掩码、网关和DNS信息，再选择Done执行下一步。

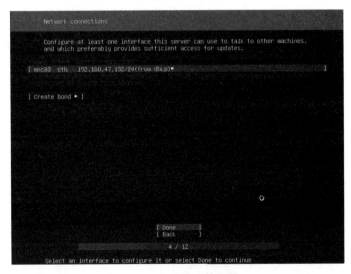

图 1-13　网络连接配置

5. 配置网络代理

配置网络代理界面如图1-14所示，下面对界面进行解释。

Configure proxy（配置代理）

If this system requires a proxy to connect to the internet, enter its details here.（如果此系统需要代理连接到Internet，请在此处输入其详细信息。）

If you need to use a HTTP proxy to access the outside world, enter the proxy information here.Otherwise, leave this blank.（如果需要使用HTTP代理访问外部世界，请在此处输入代理信息。否则，请将此项留空。）

The proxy information should be given in the standard form of "http://[[user][:pass]@]host[:port]/".（代理信息应以"http://[[user][:pass]@]host[:port]/"的标

准格式提供。)

这里直接选择Done执行下一步。

6. 配置Ubuntu镜像源

配置Ubuntu镜像源界面如图1-15所示，下面对界面进行解释。

Configure Ubuntu archive mirror（配置存档镜像源）

If you use an alternative mirror for Ubuntu, enter its details here.（如果为Ubuntu使用备用镜像，请在此处输入其详细信息。)

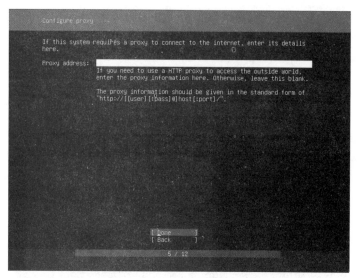

图1-14 配置网络代理

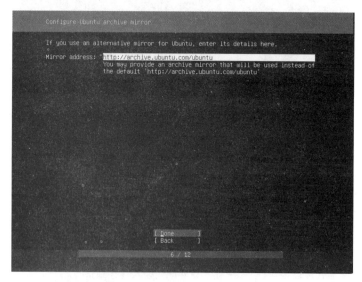

图1-15 配置Ubuntu镜像源

Mirror address（镜像地址）

You may provide an archive mirror that will be used instead of the default 'http://archive.ubuntu.com/ubuntu'.（你可以提供一个存档镜像，而不是默认的'http:

//archive.ubuntu.com/ubuntu'.)

说明：这个信息是修改apt-get源，可以在此处直接修改为清华大学、网易、阿里云等国内的源，这样在安装软件时速度会快很多。本任务采用安装完成后进行修改，所以这里直接选择Done执行下一步。

7. 配置文件系统（Filesystem setup）

配置文件系统界面如图1-16所示，下面对界面进行解释。

Filesystem setup（配置文件系统）
The installer can guide you through partitioning an entire disk either directly or using LVM, or, if you prefer, you can do it manually.（安装程序可以指导你直接或使用LVM对整个磁盘进行分区，或者，如果你愿意，你可以手动分区。）
If you choose to partition an entire disk you will still have a chance to review and modify the results.（如果选择对整个磁盘进行分区，你仍有机会查看和修改结果）。
Use An Entire Disk（使用整个磁盘）
Use An Entire Disk And Set Up LVM（使用整个磁盘并设置LVM）
Manual（手动）
Back（返回）

这里选择Manual手动分区。

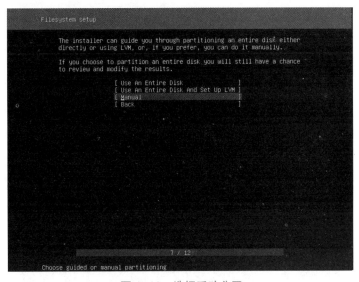

图1-16　选择手动分区

选择手动分区后，出现文件系统设置界面，如图1-17所示，下面解释界面英文内容信息。

Filesystem setup（文件系统设置）
FILE SYSTEM SUMMARY（文件系统摘要）
No disk or partitions mounted（未安装磁盘或分区）
AVAILABLE DEVICES（可用设备）

选择有效的磁盘后，选择Done，进入分区界面，如图1-18所示。
这里建立一个500 MB的/boot分区，如图1-19和图1-20所示。

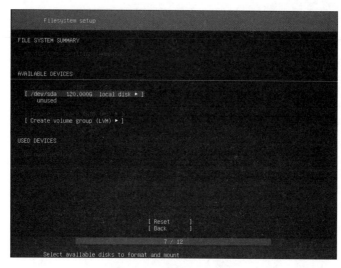

图 1-17　文件系统设置

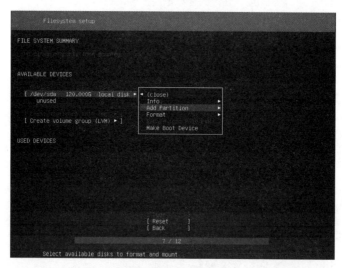

图 1-18　增加分区

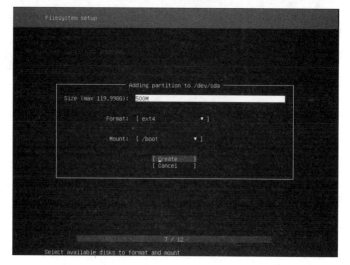

图 1-19　创建 boot 分区

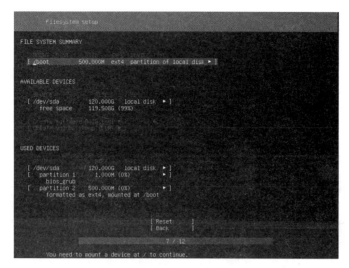

图 1-20 创建 boot 分区完成

然后，用相同的方法建立/swp分区为16 GB，因为这里设置了虚拟机的内存大小为4 GB；根分区为剩下的大小，如图1-21所示。

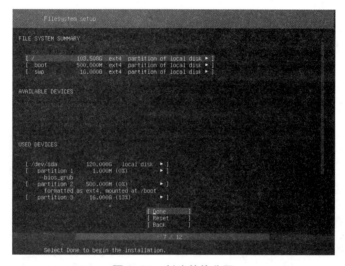

图 1-21 创建其他分区

按【Tab】键选择Done后，弹出图1-22所示的格式化警告界面，主要内容含义是：

Confirm destructive action（确认破坏性行为）

Selecting Continue below will begin the installation process and result in the loss of data on the disks selected to be formatted.（选择下面的继续将开始安装过程，并导致选定要格式化的磁盘上的数据丢失。）

You will not be able to return to this or a previous screen once the installation has started.（安装开始后，你将无法返回此屏幕或上一个屏幕。）

Are you sure you want to continue?（是否确实要继续？）

说明：这里是将划分的分区进行格式化，并且说明会丢失分区中的数据，这里选择Continue直接进入下一步。

图 1-22 格式化警告界面

8. 设置用户名

设置用户名界面如图1-23所示，下面对界面进行解释。

Enter the username and password you will use to log in to the system.（输入用于登录系统的用户名和密码。）

The name it uses when it talks to other computers.（与其他计算机对话时使用的名称）。

Pick a username [选择用户名（这是登录时输入的用户名）]

输入完成后选择Done执行下一步。

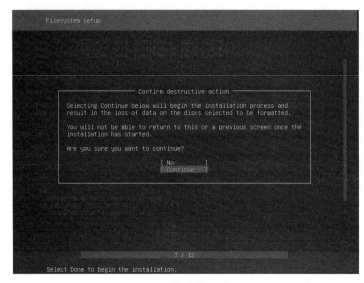

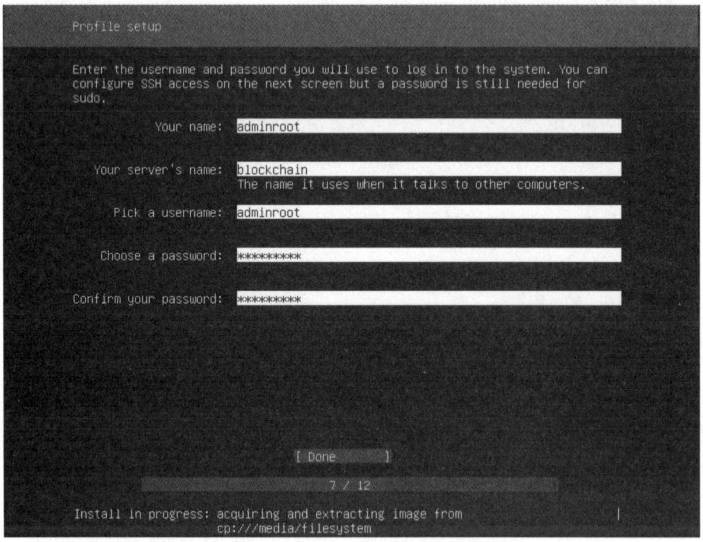

图 1-23 设置用户名

9. 安装 SSH 服务

此处要安装SSH服务，如图1-24所示。将光标移至Install OpenSSH server处按【Space】键选择，方便随后通过SSH方式登录，也可以在随后任何时间通过控制台界面安装。

项目一　认识 Docker

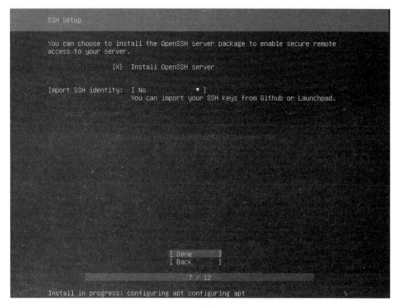

图 1-24　安装 SSH 服务

10. 安装软件包

安装软件包界面如图1-25所示。下面对界面进行解释。

These are popular snaps in server environments. Select or deselect with SPACE, press ENTER to see more details of the package, publisher and versions abailable. (这些是服务器环境中常见的软件。按【Space】键可选择或取消选择，按【Enter】键可查看可用的软件包、发布者和版本的详细信息。)

说明：这里不安装任何快照程序，因为这里的软件都是可以后面需要的时候再进行安装的，按【Tab】键选择 Done 执行下一步。

图 1-25　安装软件包界面

13

接下来是安装过程，如图1-26所示，请耐心等待，安装完成后会在[View full log]按钮下多出一个[Reboot Now]按钮。

图1-26　安装过程界面

在安装过程中可以选择View full log查看安装日志，如图1-27所示。

图1-27　安装过程的日志界面

选择Close返回，安装完成后选择Reboot Now重启Ubuntu，如图1-28所示，会跳转到下面的界面，然后按【Enter】键，系统即开始重新启动。

11. 重启登录系统查看网络配置

用自己安装时设置的用户名登录后，用ifconfig命令显示网卡配置信息，可以看到网卡ens32

的IP地址是192.168.47.132,如图1-29所示。

图1-28 安装完成后重启界面

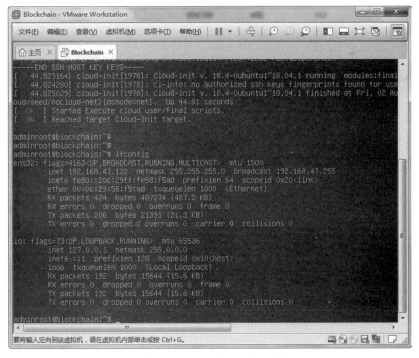

图1-29 显示网卡IP地址信息

12. 用终端登录

用终端连接,打开终端软件(PuTTY),输入主机的IP地址,单击Open按钮,如图1-30所示。

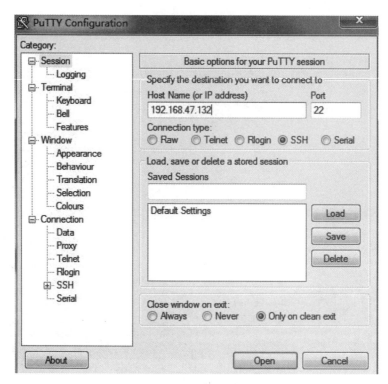

图 1-30　终端连接

输入用户名和密码进入系统，如图1-31所示。

图 1-31　终端连接完成界面

三、配置 Ubuntu 应用环境

完成Ubuntu的安装以后，下面配置Ubuntu应用环境，以方便下一步部署容器工作环境。这里主要介绍配置管理员密码、配置网卡，以及配置Ubuntu镜像源。

1. 修改 root 密码

用sudo passwd root命令修改root密码。需要强调的是，一定要记住自己设置的root密码。

```
adminroot@blockchain:~$ sudo passwd root
[sudo] password for adminroot:
Enter new UNIX password:      #第一次输入密码
Retype new UNIX password:     #第二次输入密码
passwd: password updated successfully
```

2. 修改网络设置

Ubuntu 18.04不再使用ifupdown配置网络，而改用netplan。在/etc/network/interfaces配置固定IP是无效的，重启网络的命令services network restrart或/etc/init.d/networking restart也是无效的。

（1）使用ifupdown配置网络

如果要使用之前的方式配置网络，需要重新安装ifupdown，下面举例说明：

```
sudo apt install ifupdown
```

修改配置文件/etc/network/interfaces：

```
sudo vim /etc/network/interfaces
```

配置文件修改如下：

```
iface ens160 inet static
address 210.72.92.25
gateway 210.72.92.254
netmask 255.255.255.0
dns-nameservers 8.8.8.8
```

重启网络服务使配置生效

```
sudo services network restrart
```

（2）使用netplan配置网络

Ubuntu 18.04使用netplan配置网络，其配置文件是yaml格式的（这也是未来配置文件的趋势。读者要学习yaml文件格式，在随后的学习中还有不少Docker服务配置采用yaml格式）。安装好Ubuntu 18.04之后，在/etc/netplan/目录下默认的配置文件名是50-cloud-init.yaml，通过vim命令修改它：

```
sudo vim /etc/netplan/50-cloud-init.yaml
```

本网络用自动获取方式，配置如下：

```
root@blockchain:/etc/netplan# vim 50-cloud-init.yaml
    ...   （此处省略了部分显示信息）
network:
    ethernets:
        ens32:
            dhcp4: true
    version: 2
```

下面举一个其他静态配置的例子，注意观察缩进格式，yaml文件格式有严格的缩进要求。

配置文件修改如下：

```
network:
    ethernets:
        ens160:
            addresses:
                - 210.72.92.28/24        # IP及掩码
            gateway4: 210.72.92.254      # 网关
            nameservers:
                addresses:
                    - 8.8.8.8            # DNS
    version: 2
```

无论是ifupdown还是netplan，配置的思路都是一致的，在配置文件里面按照规则填入IP、掩码、网关、DNS等信息。yaml是层次结构，需要缩进。冒号(:)表示字典，连字符(-)表示列表。

重启网络服务使配置生效：

```
sudo netplan apply
```

3. 查看DNS设置

（1）修改DNS配置文件

在Ubuntu 18.04中，DNS的管理由systemd-resolved服务实现，在/etc/resovl.conf中添加DNS配置将是临时的，重启后将失效。下面举例编辑/etc/systemd/resolved.conf。

用命令vim /etc/systemd/resolved.conf编辑，在[Resolve]下的DNS处改写为自己需要的DNS服务器地址。例如：

```
[Resolve]
DNS=192.168.47.2
…       （此处省略了部分显示信息）
```

（2）重启systemd-resolved.service服务

执行systemctl restart systemd-resolved.service命令即可。

（3）查看DNS配置信息

```
root@blockchain:/etc# systemd-resolve --status
Global
          DNSSEC NTA: 10.in-addr.arpa
                      16.172.in-addr.arpa
                      …     #此处省略部分显示信息

Link 2 (ens32)
      Current Scopes: DNS
       LLMNR setting: yes
MulticastDNS setting: no
      DNSSEC setting: no
    DNSSEC supported: no
         DNS Servers: 192.168.47.2          #本地DNS信息
          DNS Domain: localdomain
```

4. 修改源（改为阿里云的源）

国内的源速度远大于国外的源，可以在安装Ubuntu 18.04时直接修改镜像源，也可以采用如下方法配置。

（1）做备份

源的配置文件在/etc/apt目录下，首先做个备份，这是个好习惯。

```
root@blockchain:/etc/apt# cp sources.list sources.list.back
```

（2）修改源

查看信息，源配置文件是sources.list，sources.list.back是刚做的备份。

```
root@blockchain:/etc/apt# ls
apt.conf.d       sources.list       sources.list.curtin.old   trusted.gpg.d
preferences.d    sources.list.back  sources.list.d
```

修改源，可以将原来的镜像源注释掉或者删除，将阿里云的源复制到文件中（如果不知道阿里云的源，可以在网上搜索）。

```
root@blockchain:/etc/apt# vim sources.list
deb http://mirrors.aliyun.com/ubuntu/bionic main restricted universe multiverse
deb-src http://mirrors.aliyun.com/ubuntu/bionic main restricted universe multiverse
deb http://mirrors.aliyun.com/ubuntu/bionic-security main restricted universe multiverse
deb-src http://mirrors.aliyun.com/ubuntu/bionic-security main restricted universe multiverse
deb http://mirrors.aliyun.com/ubuntu/bionic-updates main restricted universe multiverse
deb-src http://mirrors.aliyun.com/ubuntu/bionic-updates main restricted universe multiverse
deb http://mirrors.aliyun.com/ubuntu/bionic-backports main restricted universe multiverse
deb-src http://mirrors.aliyun.com/ubuntu/bionic-backports main restricted universe multiverse
deb http://mirrors.aliyun.com/ubuntu/bionic-proposed main restricted universe multiverse
deb-src http://mirrors.aliyun.com/ubuntu/ bionic-proposed main restricted universe multiverse
```

软件更新测试：

```
sudo apt update
sudo apt upgrade
```

至此，成功更新。

任务总结

① 安装环境采用 VMware Workstations 虚拟机实现，或者用公有云环境创建的实例安装，也可以采用私有云环境下创建的虚拟机实现。

② 安装 Ubuntu 18.04 时尽量采用英文环境。

③ 安装过程中文件系统的设置可以采用默认或者手动方式。swap 分区现在可以保留，但是在后面学习 Kuberneters 时需要关闭 swap。

④ apt 源可以直接在安装过程中设置，也可以在安装后设置。

⑤ IP 地址、子网掩码、网关、DNS 可以直接在安装过程中设置，也可以在安装后设置。

任务扩展

重新创建一台虚拟机安装 Ubuntu 18.04，要求安装过程中手动设置网段、IP 地址、子网掩码、网关和 DNS 信息，并且设置 apt 的源为阿里云的源，安装成功后测试更新是否成功。

任务二　运行第一个测试容器

任务场景

由于业务需求，安安公司想将自己信息系统的业务移植到容器（Docker）上面。技术人员已经创建了虚拟机测试环境，现在需要测试安装和配置 Docker。假设你是云计算工程师，请完成安装和配置 Docker 工作，并运行一个容器测试。

任务描述

本任务学习配置 Docker 安装运行环境和安装 Docker，运行一个容器，并测试运行结果。

任务目标

◎ 能够配置 Docker 安装环境

◎ 能够安装 Docker 软件

◎ 能够配置加速器

◎ 能够运行一个容器并测试成功

任务实施

一、配置和测试 Docker 安装环境

配置和测试 Docker 安装环境，主要是测试宿主机与互联网的连通性，删除已经存在的旧版本，对 apt 源更新，配置必备的工具软件、添加 Docker 官方 GPG key 等工作，为安装 Docker 做准备。

1. 测试网络连通性

安装好 Ubuntu 系统后，测试与互联网的连通性，保证能通过域名方式与互联网通信，这里

测试到百度的连通性。

```
root@ubuntu16:/# ping -c 2 www.baidu.com
PING www.a.shifen.com (14.215.177.38) 56(84) bytes of data.
64 bytes from 14.215.177.38: icmp_seq=1 ttl=51 time=3.32 ms
64 bytes from 14.215.177.38: icmp_seq=2 ttl=51 time=3.57 ms
--- www.a.shifen.com ping statistics ---
2 packets transmitted, 2 received, 0% packet loss, time 4009ms
rtt min/avg/max/mdev = 2.823/3.411/3.900/0.355 ms
```

2. 删除旧版本

如果宿主机中曾经安装过的Docker版本，应执行下面的命令将其清除；如果是新安装的，该步骤略过。

```
root@blockchain:/# apt-get remove docker docker-engine docker.io
containerd runcReading package lists... Done
Building dependency tree
Reading state information... Done
Package 'docker-engine' is not installed, so not removed
...    #此处省略部分信息
0 upgraded, 0 newly installed, 0 to remove and 0 not upgraded.
```

3. 更新

```
root@blockchain:/# apt-get update
Hit:1 http://mirrors.aliyun.com/ubuntu bionic InRelease
...    #此处省略部分信息
Reading package lists... Done
root@blockchain:/#
```

4. 配置安装环境和配置源

```
root@blockchain:/# apt-get install apt-transport-https ca-certificates curl gnupg-agent software-properties-common
Reading package lists... Done
Building dependency tree
   ...（此处省略了命令执行部分显示信息）
After this operation, 54.3 kB disk space will be freed.
Do you want to continue? [Y/n] y
...（此处省略了命令执行部分显示信息）
Running hooks in /etc/ca-certificates/update.d...
done.
```

解释：apt-transport-https：是一个能使apt通过HTTPS方式访问Docker的源。计算机根据一组已存储的可信密钥检查这些签名，如果缺少有效签名或者密钥不可信，则apt会拒绝下载该文件。这样可以确保安装的软件来自授权，并且未被修改或替换。

ca-certificates：用来管理和维护证书。

curl：是一个利用URL语法在命令行下工作的文件传输工具。curl支持的通信协议有FTP、FTPS、HTTP、HTTPS、TFTP、SFTP、Gopher、SCP、Telnet、DICT、FILE、LDAP、LDAPS、

IMAP、POP3.SMTP和RTSP。

gnupg-agent：GPG代理主要用作守护进程来请求和缓存密钥链的密码。例如，用作外部程序邮件客户端等。

software-properties-common：用于添加PPA源的小工具。PPA是personal package archive的缩写，即个人包档案。使用PPA，软件制作者可以轻松地发布软件，并且能够准确地升级用户的Ubuntu。用户使用PPA源可以更加方便地获得软件的最新版本。

5. 添加 Docker 官方 GPG key

```
root@blockchain:/# curl -fsSL https://download.docker.com/linux/ubuntu/gpg | sudo apt-key add -
OK
```

此处正确显示OK就成功了。

6. 将 Docker 的源添加到 /etc/apt/sources.list

```
root@blockchain:/# add-apt-repository "deb [arch=amd64] https://download.docker.com/linux/ubuntu $(lsb_release -cs) stable"
Hit:1 http://mirrors.aliyun.com/ubuntu bionic InRelease
…（此处省略了命令执行部分显示信息）
Fetched 72.3 kB in 1s (114 kB/s)
Reading package lists... Done
```

执行完毕后执行apt-get update命令进行更新。

二、安装和配置 Docker

接下来安装Docker软件，配置Docker配置文件并配置加速器。

1. 安装 Docker 软件

执行apt-get install命令安装Docker组件，docker-ce是Docker的守护进程，docker-ce-cli 提供用户命令行接口，containerd.io提供Docker守护进程与操作系统的接口。

```
root@blockchain:/# apt-get install docker-ce docker-ce-cli containerd.io
Reading package lists... Done
…（此处省略了命令执行部分显示信息）
After this operation, 390 MB of additional disk space will be used.
Do you want to continue? [Y/n] y
…（此处省略了命令执行部分显示信息）
Processing triggers for ureadahead (0.100.0-21) ...
```

至此，命令执行完毕，安装成功。

2. 修改配置文件，实现客户端监听

该步骤不是必选步骤，以后需要时再修改也是可以的。

修改 vim /etc/systemd/system/multi-user.target.wants/docker.service，在ExecStart=项目中添加"-H tcp://0.0.0.0"，允许任何客户端监听。

```
[Service]
…（此处省略了部分信息）
ExecStart=/usr/bin/dockerd -H fd:// -H tcp://0.0.0.0 --containerd=/run/containerd/containerd.sock
```

三、配置加速器

用阿里云账号登录阿里云。如果没有阿里云账号，可以用淘宝账号或者支付宝账号直接登录，然后搜索"容器镜像服务"进入管理控制台，如图1-32所示。

图 1-32　阿里云容器镜像服务界面

配置阿里云加速器，如图1-33所示，按照提示修改配置文件。

图 1-33　配置阿里云加速器

进入/etc/docker目录，编辑下面的文件：

```
root@blockchain:/etc/docker#vim daemon.json
{
  "registry-mirrors": ["https://y6akxxyg.mirror.aliyuncs.com"]
}
```

保存并重新加载Docker：

```
root@blockchain:/etc/docker# systemctl daemon-reload
root@blockchain:/etc/docker# systemctl restart docker
```

四、运行第一个容器

该步骤实现运行一个测试容器，感受Docker功能。

1. 运行一个简单的 Apache 服务

执行docker run –d –p 80:80 httpd命令。该命令中，run是运行容器的命令；-d是后台执行，不占用终端；-p 80:80是设置端口，前面的80是暴露在宿主机的端口，后面的80是HTTP容器运行提供端口；httpd是容器镜像。

```
root@blockchain:/etc/docker# docker run -d -p 80:80 httpd
Unable to find image 'httpd:latest' locally
latest: Pulling from library/httpd
f5d23c7fed46: Pull complete
...（此处省略了命令执行部分显示信息）
Digest: sha256:dc4c86bc90593c6e4c5b06872a7a363fc7d4eec99c5d6bfac881f7371adcb2c4
Status: Downloaded newer image for httpd:latest
bed228e044534b5d0da57ea847f524fab636e1ca927871a4ae0fa6607973ffe8
```

2. 客户端测试

在任意可以与Docker host宿主机连通的客户主机上运行浏览器，输入Docker host宿主机的IP地址测试HTTP服务运行情况，如图1-34所示，表示HTTP容器运行成功。

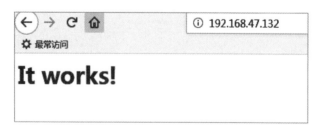

图 1-34　测试 http 服务

3. 查看 Docker 进程

在命令行中用docker ps命令查看当前正在运行的容器，从运行的结果上看，该容器正常运行。

```
root@blockchain:/# docker ps
CONTAINERID    IMAGE    COMMAND              CREATED        STATUS        PORTS                NAMES
bed228e04453   httpd    "httpd-foreground"   5 minutes ago  Up 5 minutes  0.0.0.0:80->80/tcp   practical_brattain
```

4. 测试关闭容器

至此，容器运行测试成功。对于没有用的容器可将其关闭，以免占用宿主机资源。下面测试关闭容器，需要时也可以打开容器。运行docker stop命令关闭容器。

```
root@blockchain:/# docker stop bed228e04453
bed228e04453
root@blockchain:/#
```

至此，完成此任务的Docker安装，并测试运行了一个容器。

任务总结

① 本任务首先测试到互联网的连通性，然后配置Docker安装环境，安装Docker，配置Docker运行环境，测试容器的运行。
② 加速器的配置不是必需步骤，但是可以提高下载速度。
③ 配置docker.service文件在本任务中也不是必需步骤，在以后学习中会有作用。
④ 对源的更新是必不可少的步骤，将大大提高速度。有时无法安装软件，可以尝试更新源，然后尝试再次安装。

任务扩展

尝试在本任务配置的环境中运行Nginx服务，在浏览器中访问Nginx服务。

任务三 掌握容器的基本操作

任务场景

由于业务需求，安安公司想将自己信息系统的业务移植到容器（Docker）上面。现在技术人员已经创建了一个测试环境，并运行了一个测试容器。为更进一步了解容器，需要掌握容器的基本操作，为以后进一步使用容器奠定基础。

任务描述

本任务学习容器的基本操作，包括运行容器、查看容器运行状态、停止容器运行、启动容器运行、暂停容器、重新回复容器运行、批量停止容器运行和批量删除容器等。

任务目标

◎ 能够在Ubuntu 18.04操作系统环境中启动容器
◎ 能够查看容器的运行状态，并能解释查看到的参数
◎ 能够停止容器运行和重新启动容器运行
◎ 能够暂停容器运行和重新恢复容器运行
◎ 能够批量停止容器运行和批量删除容器

任务实施

一、运行一个容器

运行一个httpd容器，端口映射到宿主机的80端口上。

root@blockchain:/# docker run -d -p 80:80 httpd
6f619f94b7ca0922541739a6cdcb40c3038df99f5239fde1d618e7589752d808
root@blockchain:/#

二、查看容器的运行

```
root@blockchain:/# docker ps
CONTAINER ID   IMAGE   COMMAND             CREATED          STATUS         PORTS                NAMES
6f619f94b7ca   httpd   "httpd-foreground"  25 seconds ago   Up 23 seconds  0.0.0.0:80->80/tcp   trusting_darwin
```

可以看到，该容器的ID是6f619f94b7ca，镜像是httpd，状态是Up，端口映射是宿主机80到容器80，该容器的名字是trusting_darwin。因为运行该容器没有指定名字，因此系统随机命名。

三、停掉容器

停掉httpd容器，这里用容器的短ID代表容器，也可以用容器的名字。

root@blockchain:/# docker stop 6f619f94b7ca
6f619f94b7ca

停掉容器后，查看容器运行的进程，可以看到已经没有httpd运行了。

```
root@blockchain:/# docker ps
CONTAINER ID   IMAGE   COMMAND   CREATED   STATUS   PORTS   NAMES
```

四、启动容器

用docker start命令启动容器，再次查看该容器，发现处于运行状态。

```
root@blockchain:/# docker start 6f619f94b7ca
6f619f94b7ca
root@blockchain:/# docker ps
CONTAINER ID   IMAGE   COMMAND             CREATED         STATUS
6f619f94b7ca   httpd   "httpd-foreground"  7 minutes ago   Up 4 seconds
   PORTS                NAMES
0.0.0.0:80->80/tcp   trusting_darwin
```

五、重新启动容器

重新启动容器，查看容器运行情况，发现正运行，运行的时间为10 s。

```
root@blockchain:/# docker restart 6f619f94b7ca
6f619f94b7ca
root@blockchain:/# docker ps
CONTAINER ID   IMAGE   COMMAND             CREATED          STATUS
6f619f94b7ca   httpd   "httpd-foreground"  10 minutes ago   Up 10 seconds
```

```
      PORTS               NAMES
0.0.0.0:80->80/tcp   trusting_darwin
```

六、暂停容器

用docker pause命令暂停容器，查看httpd进程状态处于Paused状态。

```
root@blockchain:/# docker pause 6f619f94b7ca
6f619f94b7ca
root@blockchain:/# docker ps
CONTAINER ID   IMAGE     COMMAND              CREATED         STATUS
6f619f94b7ca   httpd     "httpd-foreground"   13 minutes ago  Up 3 minutes (Paused)
      PORTS               NAMES
0.0.0.0:80->80/tcp   trusting_darwin
```

七、恢复容器

执行docker unpause命令后，查看容器运行进程，发现httpd进程状态处于正常状态。

```
root@blockchain:/# docker unpause 6f619f94b7ca
6f619f94b7ca
root@blockchain:/# docker ps
CONTAINER ID   IMAGE     COMMAND              CREATED         STATUS
6f619f94b7ca   httpd     "httpd-foreground"   17 minutes ago  Up 6 minutes
      PORTS               NAMES
0.0.0.0:80->80/tcp   trusting_darwin
```

八、删除容器

下面采用查看容器运行的状态，将退出的容器删除，并以运行变量的方式批量删除。

1. 查看当前所有容器进程

使用docker ps -a命令，查看当前所有进程，并关注容器的状态，可以看到有的容器退出了，有的处于正常运行状态。

```
root@blockchain:/# docker ps -a
CONTAINER ID         IMAGE                    COMMAND              CREATED          STATUS
PORTS                NAMES
  6f619f94b7ca       httpd                    "httpd-foreground"   20 minutes ago   Up 10 minutes
0.0.0.0:80->80/tcp   trusting_darwin
  9c8d865d8009       192.168.47.132:5000/busybox   "sh"            4 hours ago
Up 4 hours           inspiring_leavitt
  2c4a94bc1bb1       registry                 "/entrypoint.sh /etc…"  5 hours ago
Up 5 hours           0.0.0.0:5000->5000/tcp   private_registry
  #由于终端显示不全出现乱码，余同，不再一一标注
  960a16b33b75       64dbf3eac9ee             "/bin/bash"          23 hours ago
Exited (0) 23 hours ago                       sleepy_brown
  cf1527e86be0       ubuntu-with-vim-net      "/bin/bash"          24 hours ago
Exited (0) 24 hours ago                       thirsty_taussig
  898031dc4958       ubuntu                   "/bin/bash"          26 hours ago
```

```
Exited (0) 24 hours ago           elegant_blackwell
    eedc7e97944b        ubuntu         "/bin/bash"          29 hours ago
Exited (0) 26 hours ago           practical_swartz
    7fbddbb38d9a        ubuntu         "/bin/bash"          29 hours ago
Exited (0) 29 hours ago           determined_fermi
    7bf1bf3d052e        ubuntu         "/bin/bash"          31 hours ago
Exited (127) 31 hours ago         fervent_clarke
    bed228e04453        httpd          "httpd-foreground"   2 days ago
Exited (0) 2 days ago             practical_brattain
root@blockchain:/#
```

2. 删除状态是 exited 的容器

引用变量删除状态是exited的容器，这样可以实现批量删除。

```
root@blockchain:/# docker rm -v $(docker ps -aq -f status=exited)
960a16b33b75
cf1527e86be0
898031dc4958
eedc7e97944b
7fbddbb38d9a
7bf1bf3d052e
bed228e04453
root@blockchain:/#
```

查看容器，退出的容器已全部删除。

```
root@blockchain:/# docker ps -a
CONTAINER ID    IMAGE                COMMAND              CREATED         STATUS        PORTS                NAMES
6f619f94b7ca    httpd                "httpd-foreground"   23 minutes ago  Up 12 minutes 0.0.0.0:80->80/tcp   trusting_darwin
9c8d865d8009    192.168.47.132:5000/busybox  "sh"         4 hours ago     Up 4 hours                         inspiring_leavitt
2c4a94bc1bb1    registry             "/entrypoint.sh /etc…" 5 hours ago   Up 5 hours    0.0.0.0:5000->5000/tcp  private_registry
```

3. 停止所有容器

停止掉正在运行的容器，用变量方式停掉所有容器，然后查看容器运行情况。

```
root@blockchain:/# docker stop $(docker ps -aq )
6f619f94b7ca
9c8d865d8009
2c4a94bc1bb1
root@blockchain:/#
```

查看容器运行状况，发现还有几个退出状态的容器。

```
root@blockchain:/# docker ps -a
CONTAINER ID         IMAGE                                  COMMAND
CREATED              STATUS               PORTS             NAMES
```

```
    6f619f94b7ca        httpd                  "httpd-foreground"      29 minutes ago
Exited (0) 2 minutes ago              trusting_darwin
    9c8d865d8009        192.168.47.132:5000/busybox   "sh"            5 hours ago
Exited (137) 2 minutes ago            inspiring_leavitt
    2c4a94bc1bb1        registry               "/entrypoint.sh /etcâ€¦"  5 hours ago
Exited (2) 2 minutes ago              private_registry
    root@blockchain:/#
```

再次用变量的方式将退出的容器删除。

```
root@blockchain:/# docker rm -v $(docker ps -aq -f status=exited)
6f619f94b7ca
9c8d865d8009
2c4a94bc1bb1
root@blockchain:/#
```

再次查看容器进程，发现没有容器运行了。

```
root@blockchain:/# docker ps -a
CONTAINER ID        IMAGE                  COMMAND                 CREATED
STATUS              PORTS                  NAMES
root@blockchain:/#
```

任务总结

对容器的基本操作包括停止、暂停、重启、删除等，也可以根据实际情况采用变量方式批量操作。

任务扩展

在本任务的基础上，自己设计场景，尝试比较删除正在运行的容器和删除已经停止的容器的区别。

项目二 学习 Docker 镜像

【项目综述】

根据公司的业务需求,需要公司信息部门员工进一步认识 Docker 镜像,理解 Docker 镜像的分层结构,镜像的导入、导出和打包技术,以及如何创建公司内部的镜像库,并在实践中不断提升整个公司信息部门的业务水平。

【项目目标】

◎ 能够说出 Docker 镜像的分层结构
◎ 能够配置 dockerfile 文件并生成镜像
◎ 能够搭建本地镜像仓库并测试成功
◎ 能够搭建本地镜像仓库并测试成功
◎ 能够创建加密的私有仓库并测试成功
◎ 能够使用官方公共镜像 Registry
◎ 能够对镜像和容器进行打包应用

任务一 认识 Docker 镜像分层

任务场景

由于业务需求,安安公司已确定将自己信息系统的业务移植到容器(Docker)上面,由于公司运行容器需要自己构建镜像,现在需要技术人员认识 Docker 镜像的构成,理解 Docker 镜像的分层原理,为以后构建自己的镜像打好基础。

任务描述

本任务学习 Docker 镜像的分层,通过在现有镜像中添加层次,查看镜像构建历史,分析 Docker 层次,理解 Docker 镜像的分层原理。

任务目标

◎ 能够在容器中安装和运行程序
◎ 能够将容器打包生成新的镜像
◎ 能够查看容器的层次结构
◎ 能够根据容器的层次结构分析镜像构建过程,理解镜像分层原理

项目二　学习 Docker 镜像

任务实施

本任务通过下载一个Ubuntu镜像，运行Ubuntu容器，在该容器内安装vim程序命令，生成一个新的镜像，然后查看镜像生成历史，镜像的大小对比，分析镜像的分层结构。

一、下载一个 Ubuntu 镜像

用docker pull命令下载镜像。

```
root@blockchain:/# docker pull ubuntu
…（此处省略了命令执行部分显示信息）
Digest: sha256:c303f19cfe9ee92badbbbd7567bc1ca47789f79303ddcef56f77687d4744cd7a
Status: Downloaded newer image for ubuntu:latest
docker.io/library/ubuntu:latest
root@blockchain:/#
```

查看镜像。

```
root@blockchain:/# docker images
REPOSITORY      TAG         IMAGE ID        CREATED         SIZE
ubuntu          latest      3556258649b2    11 days ago     64.2MB
httpd           latest      ee39f68eb241    3 weeks ago     154MB
root@blockchain:/#
```

二、运行容器测试 Ubuntu 镜像

下面利用下载的Ubuntu镜像运行容器，测试镜像，在容器中执行ls命令，查看文件系统，然后退出容器返回宿主机。可以看到容器正常运行。

```
root@blockchain:/#docker run -itd ubuntu
root@898031dc4958:/#  ls
bin   dev   home   lib64   mnt   proc   run    srv   tmp   var
boot  etc   lib    media   opt   root   sbin   sys   usr
root@898031dc4958:/#  exit
exit
```

三、在容器中安装 vim 程序

1. 查看 vim 命令

```
root@898031dc4958:/#  vim
bash: vim: command not found
```

从命令执行的结果上看，该系统没有vim命令，下面用apt命令安装。

2. 安装 vim 程序

当前容器中没有vim命令，下面进行更新源，更新后再进行安装。

```
root@898031dc4958:/#  apt-get update
Get:1 http://security.ubuntu.com/ubuntu bionic-security InRelease [88.7 kB]
…（此处省略了部分显示信息）
Fetched 16.9 MB in 4min 10s (67.5 kB/s)
```

31

```
Reading package lists... Done
```

更新完毕后，安装vim命令。

```
root@898031dc4958:/# apt-get install -y vim
```

内容显示较多，此处省略显示信息，在容器中使用vim命令创建一个test.txt文件进行测试。

```
root@898031dc4958:/# vim test.txt
```

用ls命令查看根目录下有test.txt文件，此处用蓝色显示。

```
root@898031dc4958:/# ls
bin   dev   home  lib64  mnt   proc  run   srv   test.txt  usr
boot  etc   lib   media  opt   root  sbin  sys   tmp       var
root@898031dc4958:/#
```

四、查看镜像层次

通过对容器重新生成镜像，在新的镜像中查看生成历史，增加部分就是新增的镜像层次。

1. 查看容器进程

先查看容器进程，查看容器的ID和名称。

```
root@blockchain:/# docker ps
CONTAINER ID      IMAGE        COMMAND        CREATED         STATUS           PORTS          NAMES
898031dc4958      ubuntu       "/bin/bash"    12 minutes ago  Up 12 minutes                   elegant_blackwell
```

2. 创建新镜像

用docker commit生成新的镜像。

```
root@blockchain:/# docker commit elegant_blackwell ubuntu-with-vim
sha256:c76a1f5a595bb7d2eadc1445748657fcf7ee91a11e2f9bb72651dcbd5522b2bc
```

docker commit命令格式说明：从容器创建一个新的镜像。

```
# docker commit [OPTIONS] CONTAINER [REPOSITORY[:TAG]]
```

-a :提交的镜像作者；

-c :使用Dockerfile指令来创建镜像；

-m :提交时的说明文字；

-p :在commit时，将容器暂停。

3. 查看镜像列表

用docker images命令查看镜像列表，可以看到新生成的镜像ubuntu-with-vim。

```
root@blockchain:/# docker images
REPOSITORY         TAG       IMAGE ID       CREATED          SIZE
ubuntu-with-vim    latest    c76a1f5a595b   37 seconds ago   151MB
ubuntu             latest    3556258649b2   11 days ago      64.2MB
httpd              latest    ee39f68eb241   3 weeks ago      154MB
root@blockchain:/#
```

对比ubuntu和ubuntu-with-vim镜像的大小，可以看到ubuntu的镜像大小为64.2 MB，ubuntu-with-vim镜像的大小为151 MB。

4. 查看新镜像形成历史

用docker history命令查看ubuntu-with-vim镜像形成历史。

```
root@blockchain:/# docker history ubuntu-with-vim
IMAGE            CREATED          CREATED BY                SIZE           COMMENT
c76a1f5a595b     6 minutes ago    /bin/bash                 86.9MB
3556258649b2     11 days ago      /bin/sh-c #(nop) CMD["/bin/bash"]    0B
<missing>        11 days ago      /bin/sh-c mkdir-p/run/systemd&&echo'do…  7B
<missing>        11 days ago      /bin/sh-c set -xe && echo'#!/bin/sh'>/…  745B
<missing>        11 days ago      /bin/sh-c[-z"$(apt-get indextargets)"]   987kB
<missing>        11 days ago      /bin/sh-c#(nop)ADD file:3ddd02d976792b6c6â€¦
3.2MB
```

从生成镜像的历史可以看到，ID是3556258649b2的镜像就是原来的Ubuntu镜像，6 min前增加的部分形成新的镜像ID是c76a1f5a595b。

五、安装网络命令，形成新的镜像

下面进一步分析镜像的分层。

1. 安装 ping 命令

```
root@898031dc4958:/# ping www.baidu.com
bash: ping: command not found
root@898031dc4958:/#
```

从显示的信息中可以看出没有ping命令，下面安装。

```
root@898031dc4958:/# apt-get install inetutils-ping
…（此处省略了命令执行部分显示信息）
After this operation, 389 kB of additional disk space will be used.
Do you want to continue? [Y/n] y
…（此处省略了部分显示信息）
```

在容器中测试互联网网络，从命令运行结果可以看出，命令正常执行，达到了预期效果。

```
root@898031dc4958:/# ping www.baidu.com
PING www.a.shifen.com (183.232.231.174): 56 data bytes
64 bytes from 183.232.231.174: icmp_seq=0 ttl=127 time=7.315 ms
64 bytes from 183.232.231.174: icmp_seq=1 ttl=127 time=9.174 ms
^C--- www.a.shifen.com ping statistics ---
2 packets transmitted, 2 packets received, 0% packet loss
round-trip min/avg/max/stddev = 6.810/7.906/9.174/0.913 ms
```

2. 安装网络工具 ifconfig 命令

查看有没有ifconfig命令：

```
root@898031dc4958:/# ifconfig
bash: ifconfig: command not found
```

安装网络工具：

```
root@898031dc4958:/# apt-get install net-tools
Reading package lists... Done
...（此处省略了命令执行部分显示信息）
Setting up net-tools (1.60+git20161116.90da8a0-1ubuntu1) ...
E: Sub-process /usr/bin/dpkg exited unexpectedly
```

使用ifcongfig命令测试，容器网络配置信息如下：

```
root@898031dc4958:/# ifconfig
eth0: flags=4163<UP,BROADCAST,RUNNING,MULTICAST>  mtu 1500
        inet 172.17.0.2  netmask 255.255.0.0  broadcast 172.17.255.255
        ether 02:42:ac:11:00:02  txqueuelen 0  (Ethernet)
        RX packets 157  bytes 457597 (457.5 KB)
        RX errors 0  dropped 0  overruns 0  frame 0
        TX packets 138  bytes 8796 (8.7 KB)
        TX errors 0  dropped 0 overruns 0  carrier 0  collisions 0

lo: flags=73<UP,LOOPBACK,RUNNING>  mtu 65536
        inet 127.0.0.1  netmask 255.0.0.0
        loop  txqueuelen 1000  (Local Loopback)
        RX packets 0  bytes 0 (0.0 B)
        RX errors 0  dropped 0  overruns 0  frame 0
        TX packets 0  bytes 0 (0.0 B)
        TX errors 0  dropped 0 overruns 0  carrier 0  collisions 0
```

3. 生成新镜像

```
root@blockchain:/# docker commit elegant_blackwell ubuntu-with-vim-net
sha256:1e6280f27d5001c868a76b60423b802a3d54ac2b4001407428c687dbc01ef637
```

查看镜像列表，其中ubuntu-with-vim-net是新生成的镜像。

```
root@blockchain:/# docker images
REPOSITORY              TAG        IMAGE ID        CREATED             SIZE
ubuntu-with-vim-net     latest     1e6280f27d50    45 seconds ago      152MB
ubuntu-with-vim         latest     c76a1f5a595b    About an hour ago   151MB
ubuntu                  latest     3556258649b2    11 days ago         64.2MB
httpd                   latest     ee39f68eb241    3 weeks ago         154MB
root@blockchain:/#
```

从镜像列表上看，ubuntu-with-vim-net（1e6280f27d50）镜像安装网络工具后，比ubuntu-with-vim（c76a1f5a595b）镜像稍大。

查看当前运行的容器。

```
root@blockchain:/# docker ps
CONTAINER ID    IMAGE      COMMAND       CREATED       STATUS          PORTS       NAMES
898031dc4958    ubuntu     "/bin/bash"   2 hours ago   Up 10 minutes               elegant_blackwell
```

退出898031dc4958容器。

```
root@898031dc4958:/# exit
exit
```

root@blockchain:/#

再次查看当前运行的容器。

```
root@blockchain:/# docker ps
CONTAINER ID    IMAGE    COMMAND    CREATED    STATUS    PORTS    NAMES
root@blockchain:/#
```

可以看到当前没有运行容器了。

4. 运行新镜像容器，测试效果

```
root@blockchain:/# docker run -it ubuntu-with-vim-net
root@cf1527e86be0:/# ifconfig
eth0: flags=4163<UP,BROADCAST,RUNNING,MULTICAST>  mtu 1500
        inet 172.17.0.2  netmask 255.255.0.0  broadcast 172.17.255.255
        ether 02:42:ac:11:00:02  txqueuelen 0  (Ethernet)
        RX packets 10  bytes 836 (836.0 B)
        RX errors 0  dropped 0  overruns 0  frame 0
        TX packets 0  bytes 0 (0.0 B)
        TX errors 0  dropped 0 overruns 0  carrier 0  collisions 0

lo: flags=73<UP,LOOPBACK,RUNNING>  mtu 65536
        inet 127.0.0.1  netmask 255.0.0.0
        loop  txqueuelen 1000  (Local Loopback)
        RX packets 0  bytes 0 (0.0 B)
        RX errors 0  dropped 0  overruns 0  frame 0
        TX packets 0  bytes 0 (0.0 B)
        TX errors 0  dropped 0 overruns 0  carrier 0  collisions 0
```

显示容器网络配置信息，然后测试网络连通性。

```
root@cf1527e86be0:/# ping www.baidu.com
PING www.a.shifen.com (111.45.3.177): 56 data bytes
64 bytes from 111.45.3.177: icmp_seq=0 ttl=127 time=8.215 ms
64 bytes from 111.45.3.177: icmp_seq=1 ttl=127 time=8.130 ms
^C--- www.a.shifen.com ping statistics ---
2 packets transmitted, 2 packets received, 0% packet loss
round-trip min/avg/max/stddev = 8.130/8.969/10.563/1.127 ms
```

退出容器。

```
root@cf1527e86be0:/# exit
exit
root@blockchain:/#
```

5. 查看新生成的镜像层次

```
root@blockchain:/# docker history ubuntu-with-vim-net
IMAGE              CREATED                CREATED BY                                 SIZE        COMMENT
1e6280f27d50       14 minutes ago         /bin/bash                                  87.7MB
3556258649b2       11 days ago            bin/sh-c #(nop) CMD["/bin/bash"]           0B
```

```
<missing>      11 days ago  /bin/sh-c mkdir -p/run/systemd&&echo'do€¦   7B
<missing>      11 days ago  /bin/sh-c set -xe &&echo'#!/bin/sh'>/€¦    745B
<missing>      11 days ago  /bin/sh-c[-z"$(apt-get indextargets)"]    987kB
<missing>      11 days ago  /bin/sh-c#(nop)ADD file:3ddd02d976792b6c6€¦  63.2MB
```

从生成镜像的历史可以看到，ID是3556258649b2的镜像就是原来的Ubuntu镜像，14 min前增加的部分形成新的镜像ID是1e6280f27d50。

从以上分析可以看出，容器中的镜像是分层的，如图2-1所示。

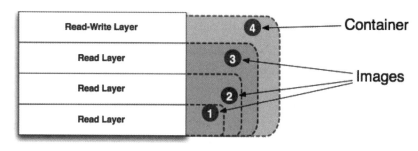

图 2-1　镜像分层的可读性

一个Docker镜像由多个可读的镜像层组成，然后运行的容器会在这个Docker的镜像上面多加一层可写的容器层，任何的对文件的更改都只存在于此容器层，因此任何对容器的操作均不会影响到镜像。

容器如何获取镜像层文件而又不影响到镜像层呢？Docker是这样实现的、如果需要获取某个文件，那么容器层会从上到下去下一层的镜像层去获取文件，如果该层文件不存在，那么就会去下一镜像层去寻找，直到最后一层。

用户面向的是一个叠加后的文件系统，下面是镜像层，最上面是容器层，如图2-2所示。

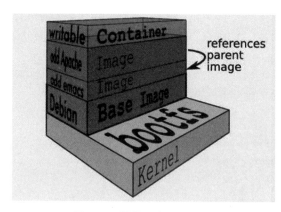

图 2-2　镜像和容器层关系

任何对于文件的操作都会记录在容器层。例如，修改文件时容器层会把在镜像层找到的文件复制到容器层然后进行修改，删除文件时则会在容器层内记录删除文件的记录。

这样做的好处是，基本上每个软件都是基于某个镜像去运行的，因此，如果某个底层环境出了问题，就不需要去修改全部基于该镜像的软件的镜像，只需要修改底层环境的镜像即可。

这样做可以使得其他相同环境的软件镜像都共同去享用同一个环境镜像，而不需要每个软件镜像要去创建一个底层环境。

任务扩展

测试Docker官方Alpine镜像，安装vim软件，然后生成镜像，查看镜像的分层。

任务二　利用 dockerfile 生成镜像

任务场景

由于业务需求，安安公司已确定将自己信息系统的业务移植到容器（Docker）上面。由于公司运行容器需要自己构建镜像，现在需要技术人员利用Docker技术创建适合自己的镜像。

任务描述

本任务学习dockerfile文件的语法格式，dockerfile的命令，利用dockerfile完成创建公司自己的镜像。

任务目标

◎ 能够描述dockerfile功能
◎ 能够描述常用dockerfile命令功能和使用方法
◎ 能够根据需求编写dockerfile文件
◎ 能够利用dockerfile文件生成镜像
◎ 进一步认识和理解镜像分层

任务实施

本任务通过创建dockerfile文件，利用dockerfile生成镜像，查看镜像的生成历史过程。

一、创建文件夹

创建存放dockerfile的文件夹，随后生成镜像时，利用该文件夹下的dockerfile文件。

```
root@blockchain:/# mkdir dockerfile
root@blockchain:/# cd dockerfile/
root@blockchain:/dockerfile#
```

二、创建 dockerfile

使用vim编辑dockerfile文件。

```
root@blockchain:/dockerfile# vim dockerfile
```

输入如下内容：

```
FROM ubuntu
```

```
RUN apt-get update && apt-get install -y  net-tools inetutils-ping
```

该dockerfile文件中共有两条指令。第一条指令FROM 后面跟上原始的ubuntu镜像，说明新构建的镜像采用的原始镜像是ubuntu；第二条指令是先进行更新，然后安装net-tools和 inetutils-ping工具。

查看创建的dockerfile文件。

```
root@blockchain:/dockerfile# ls
dockerfile
```

三、创建新镜像

运用docker build命令创建新的镜像。

```
root@blockchain:/dockerfile# docker build -t ubuntu-nettools-from-dockerfile.
Sending build context to Docker daemon  2.048kB
Step 1/2 : FROM ubuntu
 ---> 3556258649b2
Step 2/2 : RUN apt-get update && apt-get install -y  net-tools inetutils-ping
 ---> Running in 86f930b25ceb
Get:1 http://archive.ubuntu.com/ubuntu bionic InRelease [242 kB]
……（此处省略了部分显示信息）
Setting up netbase (5.4) ...
Setting up inetutils-ping (2:1.9.4-3) ...
Removing intermediate container 86f930b25ceb
 ---> 64dbf3eac9ee
Successfully built 64dbf3eac9ee
Successfully tagged ubuntu-nettools-from-dockerfile:latest
```

从命令执行过程可以看出，生成镜像分为两个步骤。第一个步骤"Step 1/2 : FROM ubuntu"，执行dockerfile文件中的第一条指令，使用的镜像ID是3556258649b2；第二个步骤"Step 2/2 : RUN apt-get update && apt-get install -y net-tools inetutils-ping"，执行dockerfile文件中的第二条指令，生成了临时的容器，ID是86f930b25ceb，并在此容器中安装net-tools和 inetutils-ping工具。从最后4行显示信息可以看出，删除了临时容器86f930b25ceb，创建了新的镜像64dbf3eac9ee，名称是"ubuntu-nettools-from-dockerfile:latest"。

四、查看镜像

用docker images命令查看当前的镜像列表，可以看到通过docker build命令生成了一个新的镜像，名字为ubuntu-nettools-from-dockerfile，下面用蓝色显示。

```
root@blockchain:/dockerfile# docker images
REPOSITORY                         TAG       IMAGE ID       CREATED          SIZE
ubuntu-nettools-from-dockerfile    latest    64dbf3eac9ee   2 minutes ago    92.6MB
ubuntu-with-vim-net                latest    1e6280f27d50   45 minutes ago   152MB
ubuntu-with-vim                    latest    c76a1f5a595b   2 hours ago      151MB
ubuntu                             latest    3556258649b2   11 days ago      64.2MB
httpd                              latest    ee39f68eb241   3 weeks ago      154MB
```

五、查看镜像分层

使用docker history命令查看镜像分层，创建（CREATED）8 minutes ago的是刚添加的一层，可以看到CREATED BY具体执行的命令。

```
root@blockchain:/dockerfile# docker history 64dbf3eac9ee
IMAGE           CREATED        CREATED BY                                       SIZE      COMMENT
64dbf3eac9ee    8 minutes ago  /bin/sh -c apt-get update && apt-get install…    8.4MB
3556258649b2    11 days ago    /bin/sh -c #(nop)  CMD ["/bin/bash"]             0B
<missing>       11 days ago    /bin/sh -c mkdir -p /run/systemd && echo 'do…    7B
<missing>       11 days ago    /bin/sh -c set -xe   && echo '#!/bin/sh' > /…    745B
<missing>       11 days ago    /bin/sh -c [ -z "$(apt-get indextargets)" ]      987kB
<missing>       11 days ago    /bin/sh -c #(nop) ADD file:3ddd02d976792b6c6…    63.2MB
```

六、运行新镜像测试安装的工具

使用docker run命令利用新的镜像64dbf3eac9ee运行容器，运行 ifconfig和ping命令，执行正常，达到测试目的。

```
root@blockchain:/dockerfile# docker run -it 64dbf3eac9ee
root@960a16b33b75:/# ls
bin   dev   home   lib64   mnt   proc   run   srv   tmp   var
boot  etc   lib    media   opt   root   sbin  sys   usr
root@960a16b33b75:/# ifconfig
eth0: flags=4163<UP,BROADCAST,RUNNING,MULTICAST>  mtu 1500
        inet 172.17.0.2  netmask 255.255.0.0  broadcast 172.17.255.255
        ether 02:42:ac:11:00:02  txqueuelen 0  (Ethernet)
        RX packets 10  bytes 836 (836.0 B)
        RX errors 0  dropped 0  overruns 0  frame 0
        TX packets 0  bytes 0 (0.0 B)
        TX errors 0  dropped 0 overruns 0  carrier 0  collisions 0

lo: flags=73<UP,LOOPBACK,RUNNING>  mtu 65536
        inet 127.0.0.1  netmask 255.0.0.0
        loop  txqueuelen 1000  (Local Loopback)
        RX packets 0  bytes 0 (0.0 B)
        RX errors 0  dropped 0  overruns 0  frame 0
        TX packets 0  bytes 0 (0.0 B)
        TX errors 0  dropped 0 overruns 0  carrier 0  collisions 0

root@960a16b33b75:/#
root@960a16b33b75:/# ping www.baidu.com
PING www.a.shifen.com (183.232.231.172): 56 data bytes
64 bytes from 183.232.231.172: icmp_seq=0 ttl=127 time=10.784 ms
64 bytes from 183.232.231.172: icmp_seq=1 ttl=127 time=9.480 ms
^C--- www.a.shifen.com ping statistics ---
2 packets transmitted, 2 packets received, 0% packet loss
round-trip min/avg/max/stddev = 9.480/10.367/10.784/0.519 ms
```

至此，生成的镜像能成功运行容器，达到预期效果。

任务总结

1. dockerfile 基本功能

dockerfile是具有一定格式的文档,用户可以基于dockerfile生成自己需要的镜像,从而定制自己的容器。

dockerfile仅仅是用来制作镜像的源码文件,是构建容器过程中的指令,docker能够读取dockerfile的指定进行自动构建容器。

基于dockerfile制作镜像,每一个指令都会创建一个镜像层,即镜像都是多层叠加而成,因此,层越多,效率越低;创建镜像,层越少越好,因此能在一个指令完成的动作尽量通过一个指令定义。

2. dockerfile 镜像制作的工作逻辑

首先需要有一个制作镜像的目录,该目录下有个文件,名称必须为dockerfile。dockerfile有指定的格式,#号开头为注释,指定默认用大写字母来表示,以区分指令和参数,docker build读取dockerfile是按顺序依次dockerfile里的配置,且第一条非注释指令必须是FROM开头,表示基于哪个基础镜像来构建新镜像。可以根据已存在的任意镜像来制作新镜像。

dockerfile可以使用环境变量,用ENV来定义环境变量,变量名支持bash的变量替换。例如,${variable:-word}表示如果变量值存在,就使用原来的变量,变量为空时,就使用word的值作为变量的值;${variable:+word}表示如果变量存在且不是空值,那么变量将会被赋予为word对应的值,如果变量为空,那么依旧是空值。

dcokerignore:把文件路径写入到dockerignore,对应的路径将不会被打包到新镜像。

任务扩展

在互联网上查找有关dockerfile指令的介绍,相互提问解释指令的含义。

任务三 搭建本地镜像仓库

任务场景

由于业务需求,安安公司已确定将自己信息系统的业务移植到容器(Docker)上面。由于公司已构建了自己的镜像,为方便技术人员共同使用镜像,需要创建自己公司的镜像库。

任务描述

本任务学习构建公司自己的镜像库,并测试使用自己的镜像库下载镜像运行容器。

任务目标

◎ 能描述构建公司自己镜像库工作流程
◎ 能下载registry镜像和运行registry容器
◎ 能更改docker配置文件使用自己的镜像库
◎ 能利用公司自己的镜像库上传和下载镜像

任务实施

本任务拉取registry镜像源并运行registry容器，然后测试在本地镜像库上上传和下载镜像。

一、环境以及准备工作

1. 检查 ubuntu 18.04 网络环境

检查ubuntu 18.04网络环境是否能正常与互联网通信。对于已安装配置好的Docker环境，用sudo apt-get update命令更新apt-get源，防止下载出错。

2. 启动 Docker 并拉取 registry 镜像源

创建本地镜像，实际上是运行一个registry容器，由该容器管理本地镜像库，因此需要拉取registry镜像源，随后运行该registry容器。

```
root@blockchain:/# docker pull registry
Using default tag: latest
latest: Pulling from library/registry
…（此处省略了部分显示内容）
Digest: sha256:8004747f1e8cd820a148fb7499d71a76d45ff66bac6a29129bfdbfdc0154d146
Status: Downloaded newer image for registry:latest
docker.io/library/registry:latest
```

3. 查看下载的镜像

下载好后查看是否下载成功，下面的代码表示已有registry镜像，蓝色显示。

```
root@blockchain:/# docker images
REPOSITORY          TAG           IMAGE ID        CREATED         SIZE
…（此处省略了部分显示内容）
registry            latest        f32a97de94e1    5 months ago    25.8MB
root@blockchain:/#
```

4. 运行 registry 容器

下载后启动该容器，可以将容器内的数据映射挂载在自己指定的目录上，这里以/opt/registry为宿主机存储的目录。

```
root@blockchain:/opt# mkdir registry    //创建目录
root@blockchain:/opt# cd registry/
root@blockchain:/opt/registry# mkdir docker_registry
```

运用下面命令：docker run -d -p 5000:5000 -v ./docker_registry:/var/lib/registry --name private_registry registry   //启动容器

- -d：让容器可以后台运行
- -p：指定映射端口（前者是宿主机的端口号，后者是容器的端口号）
- -v：数据挂载（前者是宿主机的目录，后者是容器的目录）
- --name：为运行的容器命名

```
root@blockchain:/opt/registry# docker run -d -p 5000:5000 -v ./docker_registry:/var/lib/registry --name private_registry registry
2c4a94bc1bb1a56557f7e01a0090d2233bbe74069e31967ae66a5819e5824f07
root@blockchain:/opt/registry#
```

然后查看是否成功启动该容器。

```
root@blockchain:/opt/registry# docker ps
CONTAINER ID        IMAGE               COMMAND                  CREATED             STATUS              PORTS                    NAMES
2c4a94bc1bb1        registry            "/entrypoint.sh /etc…"    2 minutes ago      Up 2 minutes        0.0.0.0:5000->5000/tcp   private_registry
root@blockchain:/opt/registry#
```

注意：5000 端口一定得加。宿主机默认是访问 80 端口，如果此处不想加 5000 端口，可以在启动容器的时候以宿主机的 80 端口映射容器的 5000 端口。

5. 修改 daemon.json 配置文件

更改Docker的配置文件，增加一行"insecure-registries":["192.168.47.132:5000"，添加自己的私库地址，Docker启动时会加载。

```
root@blockchain:/# cd /etc/docker/
root@blockchain:/#:/etc/docker# vim daemon.json
```

输入：

```
{
    "registry-mirrors": [
      "https://registry.docker-cn.com",
      "https://y6akxxyg.mirror.aliyuncs.com,
      "insecure-registries":["192.168.47.132:5000"]
    ],
    "dns": ["8.8.8.8","8.8.4.4"]
}
```

修改后重启容器并开启registry服务。

```
root@blockchain:/# service docker restart              //重启容器
root@blockchain:/#docker restart private_registry      //重启registry服务
```

查看进程，确认private_registry容器正常开启。

```
root@blockchain:/# docker ps
CONTAINER ID        IMAGE               COMMAND                  CREATED             STATUS              PORTS                    NAMES
2c4a94bc1bb1        registry            "/entrypoint.sh /etc…"    9 minutes ago      Up 17 seconds       0.0.0.0:5000->5000/tcp   private_registry
```

至此，已搭建好私库，下面进行测试。

二、测试

通过下载一个镜像，然后将该镜像打标记，上传到自己的仓库中，再从自己的仓库中下载镜像，启动容器来测试本地镜像库的作用。

1. 拉取一个镜像并打 tag（以 busybox 为例，因为 busybox 比较小）

```
root@blockchain:/# docker pull busybox:latest         //拉取镜像
```

查看镜像，已下载了busybox。

```
root@blockchain:/# docker images
REPOSITORY                          TAG       IMAGE ID       CREATED          SIZE
ubuntu-nettools-from-dockerfile     latest    64dbf3eac9ee   19 hours ago     92.6MB
ubuntu-with-vim-net                 latest    1e6280f27d50   20 hours ago     152MB
ubuntu-with-vim                     latest    c76a1f5a595b   21 hours ago     151MB
ubuntu                              latest    3556258649b2   12 days ago      64.2MB
busybox                             latest    db8ee88ad75f   2 weeks ago      1.22MB
httpd                               latest    ee39f68eb241   3 weeks ago      154MB
registry                            latest    f32a97de94e1   5 months ago     25.8MB
root@blockchain:/# docker tag busybox:latest 192.168.47.132:5000/busybox
```

2. 提交已 tag 的镜像到自己的本地镜像仓库

```
root@blockchain:/# docker push 192.168.47.132:5000/busybox
The push refers to repository [192.168.47.132:5000/busybox]
0d315111b484: Pushed
latest: digest: sha256:895ab622e92e18d6b461d671081757af7dbaa3b00e3e28e1250
5af7817f73649 size: 527
root@blockchain:/#
```

3. 删除本地所有的关于 busybox 镜像并查看

```
root@blockchain:/# docker rmi busybox 192.168.47.132:5000/busybox
Untagged: busybox:latest
Untagged: busybox@sha256:9f1003c480699be56815db0f8146ad2e22efea85129b5b598
3d0e0fb52d9ab70
Untagged: 192.168.47.132:5000/busybox:latest
Untagged: 192.168.47.132:5000/busybox@sha256:895ab622e92e18d6b461d6710817
57af7dbaa3b00e3e28e12505af7817f73649
Deleted: sha256:db8ee88ad75f6bdc74663f4992a185e2722fa29573abcc1a19186cc5ec
09dceb
Deleted: sha256:0d315111b4847e8cd50514ca19657d1e8d827f4e128d172ce8b2f76a04
f3faea
root@blockchain:/#
```

用 docker images 命令查看是否还有 busybox 镜像的信息，发现已经没有 busybox 镜像了。

```
root@blockchain:/# docker images
REPOSITORY                          TAG       IMAGE ID       CREATED          SIZE
ubuntu-nettools-from-dockerfile     latest    64dbf3eac9ee   19 hours ago     92.6MB
ubuntu-with-vim-net                 latest    1e6280f27d50   20 hours ago     152MB
ubuntu-with-vim                     latest    c76a1f5a595b   21 hours ago     151MB
ubuntu                              latest    3556258649b2   12 days ago      64.2MB
httpd                               latest    ee39f68eb241   3 weeks ago      154MB
registry                            latest    f32a97de94e1   5 months ago     25.8MB
root@blockchain:/#
```

4. 从镜像仓库 pull busybox 镜像并查看

root@blockchain:/# docker pull 192.168.47.132:5000/busybox
Using default tag: latest
latest: Pulling from busybox
ee153a04d683: Pull complete
Digest: sha256:895ab622e92e18d6b461d671081757af7dbaa3b00e3e28e12505af7817f73649
Status: Downloaded newer image for 192.168.47.132:5000/busybox:latest
192.168.47.132:5000/busybox:latest
root@blockchain:/#docker images
                //查看192.168.47.132:5000/busybox镜像的信息

发现有192.168.47.132:5000/busybox镜像，下面蓝色显示部分。

root@blockchain:/# docker images

REPOSITORY	TAG	IMAGEID	CREATED	SIZE
ubuntu-nettools-from-dockerfile	latest	64dbf3eac9ee	19 hours ago	2.6MB
ubuntu-with-vim-net	latest	1e6280f27d50	20 hours ago	152MB
ubuntu-with-vim	latest	c76a1f5a595b	21 hours ago	151MB
ubuntu	latest	3556258649b2	12 days ago	64.2MB
192.168.47.132:5000/busybox	latest	db8ee88ad75f	2 weeks ago	1.22MB
httpd	latest	ee39f68eb241	3 weeks ago	154MB
registry	latest	f32a97de94e1	5 months ago	25.8MB

root@blockchain:/#

5. 运行 busybox 容器测试

root@blockchain:/# docker run -it 192.168.47.132:5000/busybox

已经进入容器中，显示目录进行查看。

/# ls
bin   dev   etc   home   proc   root   sys   tmp   usr   var

用ifconfig命令显示该容器的网络配置信息。

/# ifconfig
eth0   Link encap:Ethernet   HWaddr 02:42:AC:11:00:03
           inet addr:172.17.0.3   Bcast:172.17.255.255   Mask:255.255.0.0
           UP BROADCAST RUNNING MULTICAST   MTU:1500   Metric:1
           RX packets:7 errors:0 dropped:0 overruns:0 frame:0
           TX packets:0 errors:0 dropped:0 overruns:0 carrier:0
           collisions:0 txqueuelen:0
           RX bytes:586 (586.0 B)   TX bytes:0 (0.0 B)

lo   Link encap:Local Loopback
           inet addr:127.0.0.1   Mask:255.0.0.0
           UP LOOPBACK RUNNING   MTU:65536   Metric:1
           RX packets:0 errors:0 dropped:0 overruns:0 frame:0
           TX packets:0 errors:0 dropped:0 overruns:0 carrier:0
           collisions:0 txqueuelen:1000

```
                RX bytes:0 (0.0 B)   TX bytes:0 (0.0 B)
```

在该容器中ping互联网，测试连通性。

```
/# ping www.baidu.com
PING www.baidu.com (183.232.231.174): 56 data bytes
64 bytes from 183.232.231.174: seq=0 ttl=127 time=8.286 ms
64 bytes from 183.232.231.174: seq=2 ttl=127 time=8.569 ms
^C
--- www.baidu.com ping statistics ---
2 packets transmitted, 2 packets received, 0% packet loss
round-trip min/avg/max = 7.542/8.132/8.569 ms
/#
```

以上运行信息表示拉取成功并成功运行容器。

任务总结

①创建的公司私有仓库实际是下载registry镜像并运行registry容器，私有仓库服务是由registry容器提供的，因此，创建公司私有仓库需要先下载registry镜像，然后运行registry容器。

②使自己的私有镜像仓库生效，需要在客户端修改docker配置文件/etc/docker/daemon.json，添加"insecure-registries":["192.168.47.132:5000"]项，用来说明使用镜像库的地址和端口号。

③测试时，将本地已有的镜像上传到公司私有镜像仓库时需要重新镜像打上tag，用于区分不同的镜像。

任务扩展

测试下载一个Nginx镜像，重新打上tag，然后上传到公司私有仓库中，测试从公司私有仓库中下载Nginx镜像并运行容器，测试容器提供的Web服务。

任务四　创建加密的私有仓库

任务场景

由于业务需求，安安公司已确定将自己信息系统的业务移植到容器（Docker）上面。由于公司已构建了自己镜像库，但是任何人均可下载使用，为了安全性，需要对下载使用公司镜像库的用户进行认证，这时，需要加密的公司镜像库。本任务构建公司自己的加密镜像库。

任务描述

本任务学习构建公司自己的加密镜像库，并测试登录加密的镜像库，上传镜像到解密的镜像库。

任务目标

◎ 能够描述构建公司自己加密镜像库创建工作流程

◎ 能够配置私有加密镜像库
◎ 能够更改Docker配置文件使用自己的加密镜像库
◎ 能够利用公司自己的加密镜像库上传和下载镜像

任务实施

创建加密的私有仓库运用到公钥加密算法，需要用openssl创建证书、公钥和私钥，还要创建访问的仓库的用户。启动仓库容器，需要指明存放用户的路径及证书与私钥的存放路径。

一、创建文件夹

创建文件夹，用来存放用户信息及私钥和证书。

```
root@blockchain:/# mkdir mytest
root@blockchain:/# cd mytest/
root@blockchain:/mytest# mkdir registry && cd registry && mkdir certs && cd certs
```

二、生成私钥和证书

1. 创建临时文件

```
root@blockchain:/mytest/registry/certs# touch /root/.rnd
```

2. 生成私钥和证书

```
root@blockchain:/mytest/registry/certs# openssl req -x509 -days 3650 -subj '/CN=192.168.47.132:5000/' -nodes -newkey rsa:2048 -keyout blockchain.key-out blockchain.crt
    Generating a RSA private key
    ........................................+++++
    ..........+++++
    writing new private key to 'blockchain.key'
    -----
```

命令解释：

req：证书申请。

-subj：指定证书请求的用户信息；如果不指定，则要求输入。

-nodes：不对私钥加密。

-newkey rsa:bits 用于生成新的rsa密钥以及证书请求；默认名称为privkey.pem。

-new: 生成新证书签署请求。

-x509: 专用于CA生成自签证书。

-key: 生成请求时用到的私钥文件。

-days n：证书的有效期限。

-out /PATH/TO/SOMECERTFILE: 证书的保存路径。

三、查看生成的秘钥文件

```
root@blockchain:/mytest/registry/certs# ls
blockchain.crt   blockchain.key
```

四、创建加密仓库访问的用户名和密码

root@blockchain:/mytest/registry# docker run --entrypoint htpasswd registry -Bbn registryman 123456 > auth/htpasswd
root@blockchain:/mytest/registry# cd auth
root@blockchain:/mytest/registry/auth# ls
htpasswd

命令完成后在/mytest/registry/auth文件夹下产生htpasswd用户文件。

五、启动并查看仓库容器

1. 启动仓库文件

root@blockchain:/mytest/registry# docker run -d -p 5000:5000 --restart=always --name registry \
  -v /mnt/registry:/var/lib/registry \
  -v /mytest/registry/auth:/auth \
  -e "REGISTRY_AUTH=htpasswd" \
  -e REGISTRY_AUTH_HTPASSWD_PATH=/auth/htpasswd \
  -e "REGISTRY_AUTH_HTPASSWD_REALM=Registry Realm" \
  -v /mytest/registry/certs:/certs \
  -e REGISTRY_HTTP_TLS_CERTIFICATE=certs/blockchain.crt \
  -e REGISTRY_HTTP_TLS_KEY=/certs/blockchain.key registry
8b7066834d03e19c760f5ae61a5f0c33ceb12b621dacacb83f1cb755cca2cd45

参数解释：
-v 挂载目录，将宿主机库文件、证书秘钥文件挂载到容器中。
-e "REGISTRY_AUTH=htpasswd"：配置环境变量，指明用户认证文件。
-e REGISTRY_AUTH_HTPASSWD_PATH=/auth/htpasswd：配置环境变量，指明用户认证文件路径。
-e "REGISTRY_AUTH_HTPASSWD_REALM=Registry Realm"：配置环境变量，指明用户名。
-e REGISTRY_HTTP_TLS_CERTIFICATE：设置环境变量告诉容器证书的位置。
-e REGISTRY_HTTP_TLS_KEY：设置环境变量告诉容器私钥的位置。

2. 查看容器运行状况

root@blockchain:/mytest/registry# docker ps -a
CONTAINER ID   IMAGE   COMMAND   CREATED   STATUS   PORTS   NAMES
8b7066834d03   registry   "/entrypoint.sh /etc…"   4 seconds ago   Up 2 seconds   0.0.0.0:5000->5000/tcp   registry

六、配置 docker registry 访问接口

1. 查看生成的证书文件夹

root@blockchain:/mytest/registry# ls
auth certs

2. 创建宿主机证书存放目录

root@blockchain:/mytest/registry#mkdir -p /ect/docker/certs.d/192.168.47.132:5000

root@blockchain:/mytest/registry# cp certs/blockchain.crt /etc/docker/certs.d/192.168.47.132:5000/

七、登记私有仓库

将私有仓库docker registry在daemon.json文件中登记（此步骤如果以前曾创建过私库可省略）。

root@blockchain:/mytest/registry# vim /etc/docker/daemon.json

然后重启docker服务。

root@blockchain:/mytest/registry#systemctl reload docker
root@blockchain:/mytest/registry#systemctl restart docker

八、测试

1. 下载 hello-world 镜像

```
root@blockchain:/mytest/registry# docker pull hello-world
Using default tag: latest
latest: Pulling from library/hello-world
1b930d010525: Pull complete
Digest: sha256:fc6a51919cfeb2e6763f62b6d9e8815acbf7cd2e476ea353743570610737b752
Status: Downloaded newer image for hello-world:latest
docker.io/library/hello-world:latest
```

2. 查看镜像

```
root@blockchain:/mytest/registry# docker images
REPOSITORY          TAG         IMAGE ID         CREATED           SIZE
...
hello-world         latest      fce289e99eb9     14 months ago     1.84kB
```

该hello-world就是新下载的镜像，此处用省略号省略了命令显示的其他镜像。

3. 标记新的镜像

```
root@blockchain:/mytest/registry# docker tag hello-world 192.168.47.132:5000/myhelloworld
root@blockchain:/mytest/registry# docker images
REPOSITORY                          TAG         IMAGE ID         CREATED           SIZE
...
192.168.47.132:5000/myhelloworld    latest      fce289e99eb9     14 months ago     1.84kB
hello-world                         latest      fce289e99eb9     14 months ago     1.84kB
```

该192.168.47.132:5000/myhelloworld就是新标记的镜像，该命令显示结果还有其他镜像，此处用省略号省略了。

4. 测试上传镜像到私库

root@blockchain:/mytest/registry# docker push 192.168.47.132:5000/myhelloworld

项目二　学习 Docker 镜像

```
The push refers to repository [192.168.47.132:5000/myhelloworld]
af0b15c8625b: Preparing 
no basic auth credentials
```

从显示信息no basic auth credentials可知需要认证，没有上传成功。

5. 登录服务器

用docker login命令登录镜像仓库服务器，输入用户名和密码，可以看到登录成功"Login Succeeded"信息。

```
root@blockchain:/mytest/registry# docker login 192.168.47.132:5000
Username: registryman             #输入登录名
Password:
WARNING! Your password will be stored unencrypted in /root/.docker/config.
json.
Configure a credential helper to remove this warning. See
https://docs.docker.com/engine/reference/commandline/login/#credentials-
store

Login Succeeded                   #登录成功
root@blockchain:/mytest/registry#
```

6. 再次推送

```
root@blockchain:/mytest/registry# docker push 192.168.47.132:5000/
myhelloworld
The push refers to repository [192.168.47.132:5000/myhelloworld]
af0b15c8625b: Pushed              #推送成功
latest: digest: sha256:92c7f9c92844bbbb5d0a101b22f7c2a7949e40f8ea90c8b3bc3
96879d95e899a size: 524
```

从命令运行结果可以看出已成功推送。

7. 查看推送的目的地

```
root@blockchain:/mytest/registry# ls /mnt/registry/docker/registry/v2/
repositories
myhelloworld
```

该目录中有个推送的myhelloworld镜像。

任务总结

①创建加密的私有仓库的目的是对使用该镜像仓库的用户进行认证，使其经过许可才能使用镜像，以保证公司内部数据的安全。

②使用加密的私有仓库中需要认证，生成证书，需要使用 openssl命令。该命令功能非常强大，本任务仅使用了生成证书的功能，读者可以根据需要了解该命令的其他功能。

③使用加密的私有仓库需要认证，通过登录的方式进行，登录失败将无法使用，因此需要妥善保管好秘钥密码。

任务扩展

测试HTTP镜像，将镜像上传到加密的私有仓库，然后下载运行容器测试。

任务五　使用官方公共镜像 Registry

任务场景

由于业务需求，安安公司已将自己信息系统的业务移植到容器（Docker）上面。由于公司已构建了自己的镜像库，为了技术分享，需要将自己的镜像分享到公共镜像库中。

任务描述

本任务学习怎样使用公共镜像库。

任务目标

◎ 能够在Docker官网生申请账号
◎ 能够将本地的镜像上传到官方公共镜像库中
◎ 能够在官方公共镜像库中查看镜像信息
◎ 能够从官方公共镜像库中下载使用

任务实施

本任务首先在官网上注册用户，然后用自己注册的用户登录，上传镜像到公共镜像库，并下载测试。具体步骤如下：

① 在Docker官网（https://hub.docker.com）注册账号，单击主页右上角的Sign Up进入注册页面，如图2-3所示。

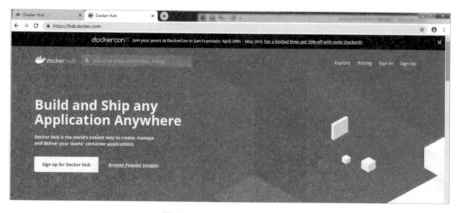

图 2-3　Docker 官网主页

② 输入账号、密码和电子邮箱等信息，并勾选下方所有的复选框，同意注册条款，如图2-4所示。

项目二　学习 Docker 镜像

图 2-4　注册页面

这样就可以创建一个账号，并可以在电子邮箱中收到激活链接。

③ 在客户端输入账号docker login –u jianqing2019（jianqing2019是笔者在Docker官网注册的账号，读者替换成自己的账号即可）及密码，登录成功后显示login succeeded，即登录成功。

```
root@blockchain:/# docker login -u jianqing2019
Password:
WARNING! Your password will be stored unencrypted in /root/.docker/config.json.
Configure a credential helper to remove this warning. See
https://docs.docker.com/engine/reference/commandline/login/#credentials-store

Login Succeeded      #显示成功
```

④ 修改镜像的repository，使之Docker hub官网账号匹配。为了区分不同用户的镜像，镜像要包括用户名（示范用户名是jianqing2019）

```
root@blockchain:/# docker tag httpd jianqing2019/httpd:v1
```

该httpd镜像是运行第一个容器时下载的，现在重新进行标记tag。查看镜像，发现有新的jianqing2019/httpd:v1镜像。

```
root@blockchain:/# docker images
REPOSITORY                    TAG       IMAGE ID       CREATED        SIZE
...
192.168.47.132:5000/busybox   latest    db8ee88ad75f   2 weeks ago    1.22MB
httpd                         latest    ee39f68eb241   3 weeks ago    154MB
jianqing2019/httpd            v1        ee39f68eb241   3 weeks ago    154MB
registry                      latest    f32a97de94e1   5 months ago   25.8MB
```

注意：上面蓝色显示的是刚才 tag 的镜像，官方自己维护的镜像没有用户名，如 httpd。

⑤ 将镜像上传到官网。通过docker push将镜像上传到docker hub。

```
root@blockchain:/# docker push jianqing2019/httpd:v1
```

51

```
The push refers to repository [docker.io/jianqing2019/httpd]
...          （此处省略了部分显示信息）
v1: digest: sha256:f2179b693cfb49baa6e7500171deea7bef755338bf165b39aedacf2
b4ae28455 size: 1367
```

⑥ 在官网查看自己上传的镜像。首先登录官网查看镜像，在 https://hub.docker.com/ 输入账号及密码登录，在公共Public Repository中可以看到自己上传的镜像，如图2-5所示。

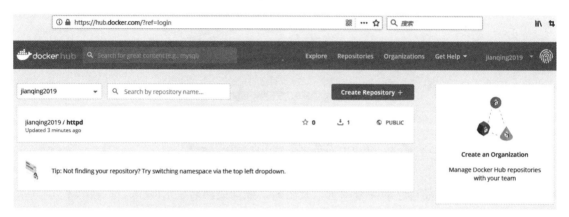

图 2-5　在官网上查看上传的镜像

还可以详细查看镜像信息，单击jianqing2019/httpd镜像，可以看到该镜像基本信息，如该镜像的tag，上传时间和执行的命令等，如图2-6所示。

图 2-6　查看上传镜像的详细信息

单击Tags，可以查看Tags信息，如图2-7所示。

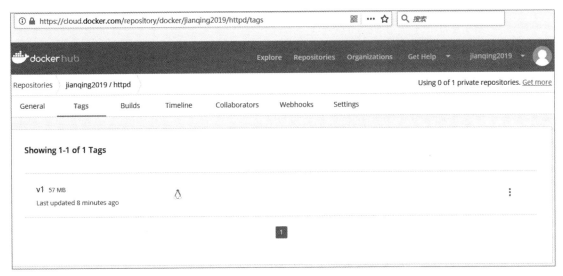

图 2-7　查看 Tags 信息

⑦ 下载测试。该镜像可被其他docker host下载使用。

```
root@blockchain:/# docker pull  jianqing2019/httpd:v1
v1: Pulling from jianqing2019/httpd
Digest: sha256:f2179b693cfb49baa6e7500171deea7bef755338bf165b39aedacf2b4
ae28455
Status: Image is up to date for jianqing2019/httpd:v1
docker.io/jianqing2019/httpd:v1
```

至此，使用官方公共镜像库测试成功。

任务总结

①使用官方公共镜像库的优势是可以实现用户不依赖本地镜像，不论在何地都可以使用自己上传的镜像，也可以不同用户在不同的客户端上使用该镜像，起到共享的作用。

②使用官方公共镜像库需要申请官方账号，使用时需要登录认证。

任务扩展

自己tag一个Nginx镜像，然后上传到官方公共镜像，查看上传的Nginx镜像的信息。

任务六　对镜像和容器进行打包

任务场景

由于业务需求，安安公司已将自己信息系统的业务移植到容器（docker）上面。为方便对镜像的管理，需要技术人员能对容器和镜像进行打包，以便于对镜像进行管理工作。

任务描述

本任务学习怎样对镜像和容器进行打包和导入打包文件。

任务目标

◎ 能够对运行的容器打包成镜像打包文件
◎ 能够对容器打包的文件导入成镜像
◎ 能够对现有的镜像打包成镜像打包文件
◎ 能够将镜像打包文件导入成镜像

任务实施

一、准备工作

Docker镜像保存在哪儿了？容器的数据默认存放在/var/lib/docker目录下，其中镜像信息存放在/var/lib/docker/image文件夹中，其哈希值在"/var/lib/docker/image/overlay2/imagedb/content/sha256"中。例如，进入该目录可以看到已经下载或者制作的镜像对应哈希值信息：

```
root@blockchain:/var/lib/docker/image/overlay2/imagedb/content/sha256# ls
0e240fbb08cd3975544a3a01bc92e858769472d227c485d3db141b93d58e76f7
1e6280f27d5001c868a76b60423b802a3d54ac2b4001407428c687dbc01ef637
3556258649b2ef23a41812be17377d32f568ed9f45150a26466d2ea26d926c32
64dbf3eac9eefe1edfd18db31435dc937e2672bc2104ded0dfaaba9e45ad5d40
6678c7c2e56c970388f8d5a398aa30f2ab60e85f20165e101053c3d3a11e6663
6ef285e58e332e6424e1c71c87113deb937f33dbb4961d483b0a126600bc45c9
708bc6af7e5e539bdb59707bbf1053cc2166622f5e1b17666f0ba5829ca6aaea
c76a1f5a595bb7d2eadc1445748657fcf7ee91a11e2f9bb72651dcbd5522b2bc
db8ee88ad75f6bdc74663f4992a185e2722fa29573abcc1a19186cc5ec09dceb
e12ed464e94177011095b38d4fd1815820f225db12cdec16556688ff2eea594a
ee39f68eb241fd811887da7e21425ac657074363daa9969b9519504785f5d60d
fce289e99eb9bca977dae136fbe2a82b6b7d4c372474c9235adc1741675f587e
```

查看当前的镜像信息可以看到镜像的ID与哈希值的对应关系：

```
root@blockchain:/var/lib/docker/image/overlay2/imagedb/content/sha256# docker images
REPOSITORY                      TAG       IMAGE ID       CREATED        SIZE
nginx                           latest    6678c7c2e56c   11 days ago    127MB
registry                        latest    708bc6af7e5e   7 weeks ago    25.8MB
amazonlinux                     latest    6ef285e58e33   4 months ago   163MB
dp_container                    latest    0e240fbb08cd   7 months ago   1.22MB
ubuntu-nettools-from-dockerfile latest    64dbf3eac9ee   7 months ago   92.6MB
ubuntu-with-vim-net             latest    1e6280f27d50   7 months ago   152MB
ubuntu-with-vim                 latest    c76a1f5a595b   7 months ago   151MB
```

```
ubuntu                              latest      3556258649b2   7 months ago    64.2MB
192.168.47.132:5000/busybox         latest      db8ee88ad75f   8 months ago    1.22MB
busybox                             latest      db8ee88ad75f   8 months ago    1.22MB
httpd                               latest      ee39f68eb241   8 months ago    154MB
jianqing2019/httpd                  v1          ee39f68eb241   8 months ago    154MB
192.168.47.132:5000/myhelloworld    latest      fce289e99eb9   14 months ago   1.84kB
hello-world                         latest      fce289e99eb9   14 months ago   1.84kB
```

虽然能看到对应的信息，但是并不能将其复制到其他宿主机成为其他用户可以使用的镜像，要实现该功能，需要用容器或者镜像打包技术实现。

二、容器镜像打包文件导出和导入

1. 将容器打包成镜像打包文件

将容器打包成镜像打包文件，该镜像信息保存放在/var/lib/docker/image中，并不能复制出来供其他宿主机使用，可以通过docker export命令将正在运行的容器保存成镜像打包文件，然后再运用docker import命令将镜像打包文件导入到宿主机中成为可以使用的镜像。下面通过使用官方ubuntu镜像操作完成这个过程。

（1）删除当前宿主机中的所有镜像

为了让大家清楚，从0镜像开始操作，保证宿主机中没有Docker镜像存在，删除所有镜像，当然这个不是必要步骤，如果很清楚条理，这个步骤可以省略。另外，如果宿主机中保存有重要的镜像，千万不要执行该命令，这里删除命令只适合临时学习使用的环境。

```
root@blockchain:/# docker rmi $(docker images -q)
```

命令说明：使用docker rm命令删除镜像，命令中的镜像ID用docker images -q查询得到，通过$进行引用。

（2）查看当前宿主机中的镜像

镜像删除完后，查看镜像，当前没有镜像。

```
root@blockchain:/# docker images
REPOSITORY          TAG         IMAGE ID        CREATED         SIZE
```

（3）下载ubuntu镜像

下载一个ubuntu镜像，以ubuntu镜像为例讲解。

```
root@blockchain:/# docker pull ubuntu
Using default tag: latest
latest: Pulling from library/ubuntu
423ae2b273f4: Pull complete
...  （此处省略了部分显示信息）
Digest: sha256:04d48df82c938587820d7b6006f5071dbbffceb7ca01d2814f81857c631d44df
Status: Downloaded newer image for ubuntu:latest
docker.io/library/ubuntu:latest
```

（4）查看当前镜像

```
root@blockchain:/# docker images
REPOSITORY          TAG            IMAGE ID          CREATED         SIZE
ubuntu              latest         72300a873c2c      3 weeks ago     64.2MB
```

此时显示有一个镜像ubuntu。

（5）运行Ubuntu容器

```
root@blockchain:/# docker run -itd --name ubuntu ubuntu /bin/bash
091d03e72bbdf9282ea8c77016d5dfcc05c2497a7c88e3f70c56c5da6a60d2c2
root@blockchain:/# docker attach ubuntu
root@091d03e72bbd:/#
```

（6）修改ubuntu容器的源

修改ubuntu容器的源，将其改成阿里云的源，本命令用echo实现，也可以通过目录挂载或者dockerfile等其他方式实现。

```
root@091d03e72bbd:/# echo 'deb http://mirrors.aliyun.com/ubuntu/bionic main restricted universe multiverse
> deb-src http://mirrors.aliyun.com/ubuntu/bionic main restricted universe multiverse
> deb http://mirrors.aliyun.com/ubuntu/bionic-security main restricted universe multiverse
> deb-src http://mirrors.aliyun.com/ubuntu/bionic-security main restricted universe multiverse
> deb http://mirrors.aliyun.com/ubuntu/bionic-updates main restricted universe multiverse
> deb-src http://mirrors.aliyun.com/ubuntu/bionic-updates main restricted universe multiverse
> deb http://mirrors.aliyun.com/ubuntu/bionic-backports main restricted universe multiverse
> deb-src http://mirrors.aliyun.com/ubuntu/bionic-backports main restricted universe multiverse
> deb http://mirrors.aliyun.com/ubuntu/bionic-proposed main restricted universe multiverse
> deb-src http://mirrors.aliyun.com/ubuntu/bionic-proposed main restricted universe multiverse' > /etc/apt/sources.list
```

（7）查看/etc/apt/sources.list文件内容

查看/etc/apt/sources.list文件内容，确保修改正确，这样随后下载安装软件的速度就快了。

```
root@091d03e72bbd:/# cat /etc/apt/sources.list
…（此处省略了部分显示信息）
```

（8）更新测试

```
root@091d03e72bbd:/# apt update
Get:1 http://mirrors.aliyun.com/ubuntu bionic InRelease [242 kB]
…（此处省略了部分显示信息）
```

```
Do you want to continue? [Y/n]
Get:1 http://mirrors.aliyun.com/ubuntu bionic-proposed/main amd64 bsdutils
amd64 1:2.31.1-0.4ubuntu3.6 [60.3 kB]
…（此处省略了部分显示信息）
Processing triggers for libc-bin (2.27-3ubuntu1) …
root@091d03e72bbd:/#
```

上面命令执行过程显示内容较多，用省略号省略了部分输出内容。从更新输出的内容上看，确实从阿里云上进行更新的速度较快。

（9）将该容器保存为镜像

该容器已成为更改为国内源（阿里云源）的容器了，将其保存为镜像，方便以后使用，这里将名称保存为ubuntu-ali-source，当然名称可以根据需要进行定义。

```
root@blockchain:/# docker commit ubuntu ubuntu-ali-source
sha256:e081db142d582c091e7cbd5b6d8baae078034dfd434dcec5e4c7610358851295
```

（10）查看当前镜像

```
root@blockchain:/# docker images
REPOSITORY           TAG      IMAGE ID       CREATED          SIZE
ubuntu-ali-source    latest   e081db142d58   56 seconds ago   126MB
ubuntu               latest   72300a873c2c   3 weeks ago      64.2MB
```

（11）将该容器打包成Ubuntu镜像打包文件

下面用docker export命令将文件保存在/docker_registry文件夹中，文件名命名为ubuntu-ali-source.tar。此处的/docker_registry文件夹是作者自己创建的，用户可以根据自己的情况创建自己使用的文件夹，保存的ubuntu-ali-source.tar文件名也是可以根据自己的喜好定义。

先查看当前运行的容器：

```
root@blockchain:/docker_registry# docker ps
CONTAINER ID    IMAGE     COMMAND       CREATED            STATUS              PORTS      NAMES
091d03e72bbd    ubuntu    "/bin/bash"   About an hour ago  Up About an hour               ubuntu
```

将容器打包：

```
root@blockchain:/docker_registry# docker export -o ubuntu-ali-source.tar 091d03e72bbd
```

命令中用-o参数进行输出，091d03e72bbd是容器的ID。下面查看打包文件，可以看到在/docker_registry文件夹中的ubuntu-ali-source.tar文件信息。

```
root@blockchain:/docker_registry# ls -l
total 114952
drwxr-xr-x 3 root root      4096 Aug  5  2019 docker
-rw------- 1 root root 117702144 Mar 16 09:22 ubuntu-ali-source.tar
```

（12）将容器镜像打包文件传输到其他host上

将容器镜像打包文件传输到其他host上，可以供其他主机导入使用，也就是说，通过docker export命令将容器打包成镜像打包文件后，可以传给任何主机进行导入使用。例如，下面传输到

192.168.47.134:/home/adminroot文件夹下，这样192.168.47.134主机就可以通过docker import命令将镜像打包文件直接导入为本机的镜像使用。

```
root@blockchain:/docker_registry# scp ubuntu-ali-source.tar 192.168.47.134:/home/adminroot
The authenticity of host '192.168.47.134 (192.168.47.134)' can't be established.
ECDSA key fingerprint is SHA256:eVLqoNhDtFrH8ZhrNXnl6sJkt4liwseNQ+Hw65eHz1U.
Are you sure you want to continue connecting (yes/no)? yes
Warning: Permanently added '192.168.47.134' (ECDSA) to the list of known hosts.
root@192.168.47.134's password:
ubuntu-ali-source.tar              100%  112MB  57.8MB/s   00:01
```

2. 将容器镜像打包文件导入为镜像

可以将容器镜像打包文件导入为本地镜像，运用docker import命令导入时，保存在本机的镜像名称是可以自定义的。例如，本例将镜像名称定义为"ubuntu-ali-source:v01"。

```
root@blockchain:/docker_registry# docker import ubuntu-ali-source.tar ubuntu-ali-source:v01
sha256:6264b7cc6f3243c12e1b30209ca7907e42d5efc71f5eebcfa66988135cd26310
```

查看镜像列表，ubuntu-ali-source:v01即为导入的镜像。

```
root@blockchain:/docker_registry# docker images
REPOSITORY            TAG       IMAGE ID        CREATED          SIZE
ubuntu-ali-source     v01       6264b7cc6f32    15 seconds ago   115MB
ubuntu-ali-source     latest    e081db142d58    45 minutes ago   126MB
ubuntu                latest    72300a873c2c    3 weeks ago      64.2MB
```

3. 测试

（1）通过ubuntu-ali-source:v01运行镜像

```
root@blockchain:/docker_registry# docker run -itd --name=ubuntu_v01 ubuntu-ali-source:v01 /bin/bash
7ed0bedc5f5e95f0d9faa215734ffb0f662a80bcb2067f42d5e0a05b19239a71
```

（2）查看正在运行的容器

```
root@blockchain:/docker_registry# docker ps
CONTAINER ID    IMAGE                    COMMAND      CREATED        STATUS         PORTS    NAMES
7ed0bedc5f5e    ubuntu-ali-source:v01    "/bin/bash"  22 seconds ago Up 20 seconds           ubuntu_v01
091d03e72bbd    ubuntu                   "/bin/bash"  4 hours ago    Up 4 hours              ubuntu
root@blockchain:/docker_registry# docker attach 7ed0bedc5f5e
root@7ed0bedc5f5e:/# ls
bin  boot  dev  etc  home  lib  lib64  media  mnt  opt  proc  root  run  sbin  srv  sys  tmp  usr  var
```

（3）查看/etc/apt/sources.list文件内容

通过查看/etc/apt/sources.list文件内容可以看出，该容器正是原来运行的容器。

```
root@7ed0bedc5f5e:/# cat /etc/apt/sources.list
    deb http://mirrors.aliyun.com/ubuntu/bionic main restricted universe
multiverse
    deb-src http://mirrors.aliyun.com/ubuntu/bionic main restricted universe
multiverse
    deb http://mirrors.aliyun.com/ubuntu/bionic-security main restricted
universe multiverse
    deb-src http://mirrors.aliyun.com/ubuntu/bionic-security main restricted
universe multiverse
    deb http://mirrors.aliyun.com/ubuntu/bionic-updates main restricted
universe multiverse
    deb-src http://mirrors.aliyun.com/ubuntu/bionic-updates main restricted
universe multiverse
    deb http://mirrors.aliyun.com/ubuntu/bionic-backports main restricted
universe multiverse
    deb-src http://mirrors.aliyun.com/ubuntu/bionic-backports main restricted
universe multiverse
    deb http://mirrors.aliyun.com/ubuntu/bionic-proposed main restricted
universe multiverse
    deb-src http://mirrors.aliyun.com/ubuntu/bionic-proposed main restricted
universe multiverse
root@7ed0bedc5f5e:/#
```

(4) 删除该容器并删除对应的镜像

先查看当前镜像：

```
root@blockchain:/docker_registry# docker images
REPOSITORY            TAG       IMAGE ID        CREATED          SIZE
ubuntu-ali-source     v01       6264b7cc6f32    3 hours ago      115MB
ubuntu-ali-source     latest    e081db142d58    4 hours ago      126MB
ubuntu                latest    72300a873c2c    3 weeks ago      64.2MB
```

删除ubuntu_v01容器：

```
root@blockchain:/docker_registry# docker rm -f ubuntu_v01
ubuntu_v01
```

查看正在运行的容器，可以看到ubuntu_v01容器已经删除：

```
root@blockchain:/docker_registry# docker ps
CONTAINER ID    IMAGE     COMMAND       CREATED       STATUS        PORTS    AMES
091d03e72bbd    ubuntu    "/bin/bash"   5 hours ago   Up 5 hours             ubuntu
```

删除ubuntu-ali-source:v01镜像：

```
root@blockchain:/docker_registry# docker rmi ubuntu-ali-source:v01
Untagged: ubuntu-ali-source:v01
Deleted: sha256:6264b7cc6f3243c12e1b30209ca7907e42d5efc71f5eebcfa66988135
cd26310
Deleted: sha256:6fe571191cd02578fb310e4b79c9caff9a596341dde2ce907ac30762
```

ca230380

再次查看镜像，发现ubuntu-ali-source:v01已经不见了。

```
root@blockchain:/docker_registry# docker images
REPOSITORY           TAG         IMAGE ID         CREATED           SIZE
ubuntu-ali-source    latest      e081db142d58     4 hours ago       126MB
ubuntu               latest      72300a873c2c     3 weeks ago       64.2MB
```

通过上面步骤，完成从下载ubuntu镜像，运行容器，然后修改容器内容，将容器打包，再将镜像打包文件导入到宿主机中成为新的镜像，通过该新镜像运行容器进行验证，该导入的镜像是成功的。

三、镜像打包并导入

前面采用docker export命令将正在运行的容器打包成镜像打包文件，然后再运用docker import命令将镜像打包文件加载到宿主机中成为镜像。下面用docker save命令将镜像打包成镜像打包文件，然后再运用docker load命令将镜像打包文件导入到宿主机中成为本机的镜像。

1. 查看当前镜像

```
root@blockchain:/docker_registry# docker images
REPOSITORY           TAG         IMAGE ID         CREATED           SIZE
ubuntu-ali-source    v01         08b9e27bac90     5 minutes ago     115MB
ubuntu               latest      72300a873c2c     3 weeks ago       64.2MB
```

2. 将 ubuntu-ali-source:v01 保存为打包文件

用docker save将ubuntu-ali-source:v01镜像打包成ubuntu-ali-source-v02.tar。

```
root@blockchain:/docker_registry# docker save ubuntu-ali-source:v01 > ubuntu-ali-source-v02.tar
```

3. 查看打包后的文件

```
root@blockchain:/docker_registry# ls
docker  ubuntu-ali-source.tar  ubuntu-ali-source-v02.tar
```

4. 删除 ubuntu-ali-source:v01 镜像

```
root@blockchain:/docker_registry# docker rmi ubuntu-ali-source:v01
Untagged: ubuntu-ali-source:v01
Deleted: sha256:08b9e27bac909db7e7252714beff6068d89d745ec097a879c678d9332ea60354
Deleted: sha256:6fe571191cd02578fb310e4b79c9caff9a596341dde2ce907ac30762ca230380
```

删除后查看镜像，发现ubuntu-ali-source:v01镜像已被删除。

```
root@blockchain:/docker_registry# docker images
REPOSITORY           TAG         IMAGE ID         CREATED           SIZE
ubuntu               latest      72300a873c2c     3 weeks ago       64.2MB
```

5. 将镜像打包文件导入到宿主机

```
root@blockchain:/docker_registry# docker load -i ubuntu-ali-source-v02.tar
```

```
    6fe571191cd0: Loading layer [================================================
=====>]   117.7MB/117.7MB
    Loaded image: ubuntu-ali-source:v01
```

再次查看镜像，发现又产生了ubuntu-ali-source:v01镜像。

```
root@blockchain:/docker_registry# docker images
REPOSITORY           TAG         IMAGE ID         CREATED           SIZE
ubuntu-ali-source    v01         08b9e27bac90     5 minutes ago     115MB
ubuntu               latest      72300a873c2c     3 weeks ago       64.2MB
```

到此，完成docker save和docker load命令对镜像打包和导入的工作。

任务总结

① docker save是将现有的镜像（image）打包成打包文件，docker export是将正在运行的容器（container）打包成打包文件。

② docker load用来导入镜像是用docker save保存的镜像打包文件，是一个分层文件系统。

③ docker import导入的镜像是用docker export保存的镜像打包文件。

④ docker load不能对载入的镜像重命名，而docker import可以为镜像指定新名称。

⑤ docker export打包文件会比save的打包文件要小，原因是save的是一个分层的文件系统，export导出的只是一个linux系统的文件目录。

任务扩展

下载一个Nginx镜像，修改默认的主页文件，用docker import和docker export测试容器镜像打包的导出和导入，用docker save和docker load测试镜像打包导出和导入。

项目三　　管理容器外加数据卷

【项目综述】

根据公司的业务需求，需要将宿主机中的数据卷挂载到容器中，方便数据及时更新和共享，根据业务不同需要定制不同的挂载方式。

【项目目标】

◎ 能够叙述数据卷不同的挂载方式

◎ 能够描述数据卷不同挂载方式不同点

◎ 能够根据业务不同选择不同的数据卷挂载方式

◎ 能够根据不同的数据卷挂载方式实现数据卷的挂载

任务一　　通过宿主机目录挂载容器数据卷

任务场景

由于业务需求，安安公司已经自己信息系统的业务移植到容器（Docker）上面，但是现在需要将宿主机上的数据共享到容器中，有的数据可以长期不发生变化，有的数据需要及时更新，有的数据属于临时挂载使用。假设你是Docker工程师，需要完成该工作。

任务描述

本任务学习数据卷的挂载，根据业务的不同，需要学习bind mount、docker managed volume方式数据卷的挂载。

任务目标

◎ 能够描述和实现bind mount数据卷挂载

◎ 能够描述和实现docker managed volume数据卷挂载

任务实施

一、认识 Docker 数据卷

容器可以通过加载数据卷的方式扩展空间，可以通过bind mount将宿主机上的卷mount到容器中使用，或者通过docker managed volume实现宿主机数据挂载，也可创建volume container来实现数据卷挂载。

Docker镜像是由多个文件系统（只读层）叠加而成。当启动一个容器的时候，Docker会加载只读镜像层并在其上（即镜像栈顶部）添加一个读写层。如果运行中的容器修改了现有的一个已经存在的文件，那该文件将会从读写层下面的只读层复制到读写层，该文件的只读版本仍然存在，只是已经被读写层中该文件的副本所隐藏。当删除Docker容器，并通过该镜像重新启动时，之前的更改将会丢失（因为重新创建了容器）。在Docker中，只读层及在顶部的读写层的组合称为UnionFile System（联合文件系统）。

Docker中的数据可以存储在类似于虚拟机磁盘的介质中，在Docker中称为数据卷（Data Volume）。数据卷可以用来存储Docker应用的数据，也可以用来在Docker容器间进行数据共享。数据卷呈现给Docker容器的形式就是一个目录，支持多个容器间共享，修改也不会影响镜像。使用Docker的数据卷，类似在系统中使用mount挂载一个文件系统。

为了能够保存（持久化）数据以及共享容器间的数据，Docker提出了Volume的概念。简单来说，Volume就是目录或者文件，它可以绕过默认的联合文件系统，而以正常的文件或者目录的形式存在于宿主机上。

二、采用 bind mount 将宿主机上的卷 mount 到容器

1. 运行 http 容器

（1）查看httpd镜像

用docker images命令查看镜像列表，已经存在httpd镜像。

```
root@blockchain:~# docker images
REPOSITORY          TAG           IMAGE ID            CREATED             SIZE
…    （此处省略了部分显示信息）
httpd               latest        ee39f68eb241        4 weeks ago         154MB
registry            latest        f32a97de94e1        5 months ago        25.8MB
```

（2）运行httpd容器

```
root@blockchain:~# docker run -d -p 80:80 httpd
5e87b401f5424b2adb3d4ba8183e21e09f55d904f6b587d2654def20941205bd
root@blockchain:~#
```

（3）测试httpd容器

在浏览器中输入192.168.47.132可以访问该页面，显示页面内容"It works!"，如图3-1所示。

图 3-1　访问 httpd 容器页面

（4）查看页面文件的内容

① 用curl 192.168.47.132命令也可查看到具体的内容。

```
root@blockchain:~# curl 192.168.47.132
<html><body><h1>It works!</h1></body></html>
root@blockchain:~#
```

说明当前httpd容器页面默认的内容是"It works!"。

② 进入容器中查看文件内容。

```
root@blockchain:~# docker exec -it 5e87b401f542 /bin/sh
# pwd
/usr/local/apache2
# ls
bin  build  cgi-bin  conf  error  htdocs  icons  include  logs  modules
# cd htdocs
# ls
index.html
# cat index.html
<html><body><h1>It works!</h1></body></html>
#
```

容器内的/usr/local/apache2/index.html文件内容正是"It works!"。

2. 挂载数据卷

（1）创建页面文件

```
root@blockchain:~# ls
root@blockchain:~# mkdir htdocs
root@blockchain:~# ls
htdocs
root@blockchain:~# cd htdocs/
root@blockchain:~/htdocs# vi index.html
```

在文件中输入下面内容：<html><body><h1>my test http web!</h1></body></html>，然后保存退出。

（2）挂载网页文件

① 容器挂载宿主机的网页文件。

挂载~/htdocs/index.html文件作为容器的数据卷。挂载卷的格式是：

docker run -it -v /宿主机绝对路径:/容器内目录: 权限   镜像名

权限：ro（容器只能查看），本例的容器内目录是"/usr/local/apache2/htdocs"。

```
root@blockchain:~/htdocs# docker run  -d --name httpd_vol -p 80:80 -v ~/htdocs:/usr/local/apache2/htdocs httpd
1b3b9abf3d99057b7674c6c5faffa97b1811cf725fbf005bb55efdb489eb528d
```

该命令将宿主机的"~/htdocs"卷挂载到容器的"/usr/local/apache2/htdocs"中，并用--name参数将该容器命名为"httpd_vol"。

② 查看容器进程。

```
root@blockchain:~/htdocs# docker ps
CONTAINER ID    IMAGE        COMMAND        CREATED       STATUS       PORTS      NAMES
```

```
1b3b9abf3d99    httpd     "httpd-foreground"    5 minutes ago    Up 5 minutes
0.0.0.0:80->80/tcp    httpd_vol
```

(3) 测试查看网页内容

① 用curl 192.168.47.132命令查看网页内容。

```
root@blockchain:~/htdocs# curl 192.168.47.132
<html><body><h1>my test http web!</h1></body></html>
```

② 进入容器中查看网页内容。

```
root@blockchain:~/htdocs# docker exec -it httpd_vol /bin/sh
# pwd
/usr/local/apache2
# ls
bin  build  cgi-bin  conf  error  htdocs  icons  include  logs  modules
# cd htdocs
# cat index.html
<html><body><h1>my test http web!</h1></body></html>
```

③ 用浏览器进行访问查看网页测试结果。

用浏览器进行访问查看网页测试结果，可以看到显示的网页内容正是宿主机~/htdocs/index.html文件中的内容，如图3-2所示。

图 3-2　挂载网页测试

④ 用docker inspect命令查看挂载信息。

```
docker inspect 1b3b9abf3d99
...（此处省略了部分显示信息）
"HostConfig": {
        "Binds": [
            "/root/htdocs:/usr/local/apache2/htdocs"
        ],
...（此处省略了部分显示信息）
 "Mounts": [
        {
            "Type": "bind",
            "Source": "/root/htdocs",           #宿主机源路径
            "Destination":"/usr/local/apache2/htdocs",   #容器目的路径
```

```
                "Mode": "",
                "RW": true,
                "Propagation": "rprivate"
            }
        ],
```
…（此处省略了部分显示信息）

从以上查看网页内容和测试结果上看，当前容器已经挂接了宿主机~/htdocs/index.html文件中的内容，并且将宿主机的内容同步到容器的/usr/local/apache2/htdocs/index.html中。

也可通过docker inspect –f {{.Mounts}} 1b3b9abf3d99命令精确查找Mounts字段的内容。

（4）更改宿主机文件内容进行测试

① 更改宿主机~/htdocs/index.html文件中的内容。

将宿主机~/htdocs/index.html文件中的内容改为"my test new http web! This is the web content of my change"。

root@blockchain:~/htdocs# vi index.html

输入"my test new http web! This is the web content of my change"，然后保存退出。

② 用curl 192.168.47.132查看网页内容。

root@blockchain:~/htdocs# curl 192.168.47.132
<html><body><h1>my test new http web! This is the web content of my change</h1></body></html>
root@blockchain:~/htdocs#

③ 用浏览器访问网站。

如图3-3所示，更改网页文件内容，容器网页随机变化。

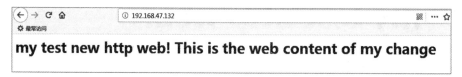

图3-3　更改挂载网页文件内容测试

从测试结果上看，对于容器没有做任何重启和网页文件内容改动，只要在宿主机中更改了卷的内容，容器中的内容就自动发生变化。

3. 测试数据卷的共享

在宿主机上的卷应该是可以供多个容器使用的，下面采用重新打开一个容器，挂载宿主机的共享卷达到卷的共享。

（1）重新开启一个httpd容器

① 挂载宿主机~/htdocs卷运行一个新的容器。

root@blockchain:~/htdocs# docker run -d --name httpd_vol_01 -p 8000:80 -v ~/htdocs:/usr/local/apache2/htdocs httpd
ae2be6a06fda19974f5d280e32114025af3d9f84aa9a53bf55f427d90be420b0
root@blockchain:~/htdocs#

该命令将容器命名为httpd_vol_01，映射的端口是8000，将宿主机的~/htdocs映射到容器的/

usr/local/apache2/htdocs。

② 查看容器进程。

```
root@blockchain:~/htdocs# docker ps
  CONTAINER ID      IMAGE              COMMAND               CREATED
STATUS                PORTS              NAMES
  ae2be6a06fda      httpd              "httpd-foreground"    10 minutes ago
Up 10 minutes         0.0.0.0:8000->80/tcp    httpd_vol_01
  1b3b9abf3d99      httpd              "httpd-foreground"    50 minutes ago
Up 50 minutes         0.0.0.0:80->80/tcp      httpd_vol
```

(2) 查看挂载卷的结果

① 用curl 192.168.47.132:8000查看网页内容。

```
root@blockchain:~/htdocs# curl 192.168.47.132:8000
<html><body><h1>my test new http web! This is the web content of my change</h1></body></html>
root@blockchain:~/htdocs# curl 192.168.47.132
<html><body><h1>my test new http web! This is the web content of my change</h1></body></html>
```

② 用浏览器访问网站。

分别在宿主机192.168.47.132的80端口和8000端口上测试网页，如图3-4和图3-5所示。

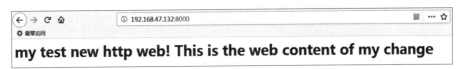

图 3-4 共享网页 8000 端口网页测试

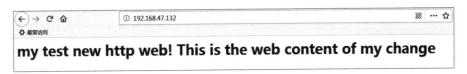

图 3-5 共享网页 80 端口网页测试

③ 用docker inspect命令查看挂载信息。

```
docker inspect ae2be6a06fda
…（此处省略了部分显示信息）
"HostConfig": {
            "Binds": [
                "/root/htdocs:/usr/local/apache2/htdocs"
            ],
…（此处省略了部分显示信息）
 "Mounts": [
            {
                "Type": "bind",
```

```
                    "Source": "/root/htdocs",
                    "Destination": "/usr/local/apache2/htdocs",
                    "Mode": "",
                    "RW": true,
                    "Propagation": "rprivate"
                }
            ],
```
… （此处省略了部分显示信息）

上面同时进行访问192.168.47.132的80端口和8000端口，两个容器正常运行，页面的内容完全一样，从测试的结果上看，宿主机上的卷是可以供多个容器使用的。

4. 测试删除容器对卷的影响

（1）删除httpd容器

① 查看运行的容器。

```
root@blockchain:~/htdocs# docker ps
CONTAINER ID        IMAGE               COMMAND                  CREATED
STATUS              PORTS               NAMES
   ae2be6a06fda     httpd               "httpd-foreground"       10 minutes ago
Up 10 minutes       0.0.0.0:8000->80/tcp   httpd_vol_01
   1b3b9abf3d99     httpd               "httpd-foreground"       50 minutes ago
Up 50 minutes       0.0.0.0:80->80/tcp     httpd_vol
```

② 停止httpd_vol_01和httpd_vol容器。

```
root@blockchain:~/htdocs# docker stop ae2be6a06fda
ae2be6a06fda
root@blockchain:~/htdocs# docker stop 1b3b9abf3d99
1b3b9abf3d99
root@blockchain:~/htdocs#
```

③ 查看退出的容器。

```
root@blockchain:~/htdocs# docker ps -aq -f status=exited
ae2be6a06fda
1b3b9abf3d99
```

④ 删除退出的容器。

```
root@blockchain:~/htdocs# docker rm -v $(docker ps -aq -f status=exited)
ae2be6a06fda
1b3b9abf3d99
```

（2）查看宿主机的文件内容

```
root@blockchain:~/htdocs# ls
index.html
root@blockchain:~/htdocs# cat index.html
<html><body><h1>my test new http web! This is the web content of my change</h1></body></html>
root@blockchain:~/htdocs#
```

从上面的测试结果看，虽然容器删除了，但是宿主机上卷的内容没有发生变化，说明删除容器对其挂载的卷没有影响。

三、通过 docker managed volume 实现宿主机数据卷挂载

docker managed volume方式挂载数据卷是由docker系统进行管理，通过在宿主机上创建文件夹并进行自动挂载。本任务通过创建容器，挂载数据卷，查找宿主机的卷位置，然后更改卷中index.html文件的内容来确认挂载数据卷的存在。

1. 运行容器并挂载数据卷

```
root@blockchain:/# docker run -d --name httpd_vol -p 80:80 -v /usr/local/apache2/htdocs httpd
86fb6ccffa9cd8dd9d350752d2352595cecb46cd02ee2a3cd75cda978f5275d2
root@blockchain:/#
```

该命令创建一个名称为httpd_vol的容器，挂载的卷mount到容器的/usr/local/apache2/htdocs目录。查看进程可以看到该容器。

```
root@blockchain:/# docker ps
CONTAINER ID        IMAGE              COMMAND               CREATED
STATUS              PORTS              NAMES
86fb6ccffa9c        httpd              "httpd-foreground"    2 minutes ago
Up 2 minutes        0.0.0.0:80->80/tcp httpd_vol
```

2. 查看网页内容

① 通过curl 192.168.47.132命令查看网页内容。

```
root@blockchain:/# curl 192.168.47.132
<html><body><h1>It works!</h1></body></html>
root@blockchain:/#
```

② 通过浏览器查看容器网站网页。这是经典的http容器测试网页，如图3-6所示。

图 3-6　http 容器测试网页

通过查看当前容器网站网页的内容是"It works!"，是该镜像默认的测试网页内容。

3. 通过 docker inspect 查看容器挂载卷宿主机的位置

```
root@blockchain:/# docker inspect 86fb6ccffa9c
... （此处省略了部分显示信息）
   "Mounts": [
        {
            "Type": "volume",
            "Name": "438c70d81149462db4c35d7e4a67ab14b70d31f566d56abc6cae976
```

```
              f111c1fee",
                        "Source": "/var/lib/docker/volumes/438c70d81149462db4c35d
7e4a67ab14b70d31f566d56abc6cae976f111c1fee/_data",
                        "Destination": "/usr/local/apache2/htdocs",
                        "Driver": "local",
                        "Mode": "",
                        "RW": true,
                        "Propagation": ""
                    }
                ],
    … （此处省略了部分显示信息）
```

从显示的内容"Source": "/var/lib/docker/volumes/438c70d81149462db4c35d7e4a67ab14b70d31f566d56abc6cae976f111c1fee/_data"，可知该容器的挂载的卷在宿主机的位置是/var/lib/docker/volumes目录下，也就是运行httpd_vol容器时，系统在宿主机的/var/lib/docker/volumes目录下创建一个名称为438c70d81149462db4c35d7e4a67ab14b70d31f566d56abc6cae976f111c1fee/_data的文件夹，挂载到容器的/usr/local/apache2/htdocs目录中。这样就实现了容器数据卷的挂载。

也可以通过docker volume命令查看卷的内容。

```
root@blockchain:/# docker volume ls
DRIVER              VOLUME NAME
local               438c70d81149462db4c35d7e4a67ab14b70d31f566d56abc6cae976
f111c1fee
```

通过docker volume inspect查看挂载点。

```
root@blockchain:/# docker volume inspect 438c70d81149462db4c35d7e4a67ab14b
70d31f566d56abc6cae976f111c1fee
[
    {
        "CreatedAt": "2019-08-12T06:35:46Z",
        "Driver": "local",
        "Labels": null,
        "Mountpoint": "/var/lib/docker/volumes/438c70d81149462db4c35d7e4a
67ab14b70d31f566d56abc6cae976f111c1fee/_data",
        "Name": "438c70d81149462db4c35d7e4a67ab14b70d31f566d56abc6cae976f
111c1fee",
        "Options": null,
        "Scope": "local"
    }
]
```

4. 修改宿主机挂载卷的网页内容

① 进入宿主机挂载卷目录。

```
root@blockchain:/# cd /var/lib/docker/volumes
root@blockchain:/var/lib/docker/volumes# ls
438c70d81149462db4c35d7e4a67ab14b70d31f566d56abc6cae976f111c1fee
```

```
metadata.db
   root@blockchain:/var/lib/docker/volumes# cd 438c70d81149462db4c35d7e4a67
ab14b70d31f566d56abc6cae976f111c1fee/
   root@blockchain:/var/lib/docker/volumes/438c70d81149462db4c35d7e4a67ab14b
70d31f566d56abc6cae976f111c1fee# ls
   _data
   root@blockchain:/var/lib/docker/volumes/438c70d81149462db4c35d7e4a67ab14b
70d31f566d56abc6cae976f111c1fee# cd _data/
   root@blockchain:/var/lib/docker/volumes/438c70d81149462db4c35d7e4a67ab14b
70d31f566d56abc6cae976f111c1fee/_data# ls
   index.html
```

② 修改index.html文件内容。

```
   root@blockchain:/var/lib/docker/volumes/438c70d81149462db4c35d7e4a67ab14b
70d31f566d56abc6cae976f111c1fee/_data# vim index.html
```

更改为"It works! This is the web content of my changed"。当然，读者可以根据自己需要更改该网页内容，修改完后保存退出。

③ 测试网页内容变化，用curl 192.168.47.132命令查看网页内容。

```
    root@blockchain:/# curl 192.168.47.132
   <html><body><h1>It works! This is the web content of my changed </h1>
</body></html>
   root@blockchain:/#
```

④ 通过浏览器查看网页内容，如图3-7所示。

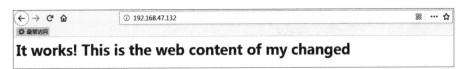

图3-7 /var/lib/docker/volumes 目录下挂载的网页

从网页显示内容可以看出，在不对httpd_vol容器做改动的情况下直接查看网页内容，网页内容已经变化为新修改后内容了，说明/var/lib/docker/volumes/438c70d81149462db4c35d7e4a67ab14b70d31f566d56abc6cae976f111c1fee/_data已经挂载到httpd_vol容器的/usr/local/apache2/htdocs目录上了。

5. 测试删除容器对卷的影响

① 停止和删除httpd_vol容器。

```
root@blockchain:/# docker stop httpd_vol
httpd_vol
root@blockchain:/# docker rm httpd_vol
httpd_vol
```

②查看卷的内容。

```
root@blockchain:/# cat /var/lib/docker/volumes/438c70d81149462db4c35d7e4a6
7ab14b70d31f566d56abc6cae976f111c1fee/_data/index.html
```

```
<html><body><h1>It works! This is the web content of my changed </h1>
</body></html>
root@blockchain:/#
```

从查看结果上看，卷仍然存在，卷中的内容没有发生变化，说明采用docker rm httpd_vol命令删除时没有带参数-v，如果带上该参数删除容器时会将卷一并删除。下面进一步测试。

③启动一个容器并加载卷。

```
root@blockchain:/# docker run  -d --name httpd_del_vol -p 80:80 -v /usr/local/apache2/htdocs httpd
88ae681d24d417c0da808a96d1c7cce58a4d1e4e0b3e3689e23c11d54d5f97e6
root@blockchain:/#
```

该命令启动一个httpd_del_vol容器，并通过docker managed volume方式挂载一个卷。

④查看卷的情况。

```
root@blockchain:/# docker volume ls
DRIVER    VOLUME NAME
local     438c70d81149462db4c35d7e4a67ab14b70d31f566d56abc6cae976f111c1fee
local     9637fe045c386e1be1de81fc2cf61be3f49c7238fb1d8608dacb39eea3f2c495
root@blockchain:/#
```

新增加的9637fe045c386e1be1de81fc2cf61be3f49c7238fb1d8608dacb39eea3f2c495就是httpd_del_vol容器挂载的卷。也可通过docker inspect命令查看，观察蓝色显示部分。

```
root@blockchain:/# docker inspect httpd_del_vol
……（此处省略了部分显示信息）
"Mounts": [
        {
            "Type": "volume",
            "Name": "9637fe045c386e1be1de81fc2cf61be3f49c7238fb1d8608dacb39eea3f2c495",
            "Source": "/var/lib/docker/volumes/9637fe045c386e1be1de81fc2cf61be3f49c7238fb1d8608dacb39eea3f2c495/_data",
            "Destination": "/usr/local/apache2/htdocs",
            "Driver": "local",
            "Mode": "",
            "RW": true,
            "Propagation": ""
        }
    ],
… （此处省略了部分显示信息）
```

⑤删除容器时一并删除卷，先停止httpd_del_vol容器。

```
root@blockchain:/# docker stop httpd_del_vol
httpd_del_vol
```

用docker rm –v httpd_del_vol命令删除httpd_del_vol容器。

```
root@blockchain:/# docker rm -v httpd_del_vol
httpd_del_vol
```

⑥查看删除后的卷情况。

```
root@blockchain:/# docker volume ls
DRIVER     VOLUME NAME
local      438c70d81149462db4c35d7e4a67ab14b70d31f566d56abc6cae976f111c1fee
root@blockchain:/#
```

可以发现，9637fe045c386e1be1de81fc2cf61be3f49c7238fb1d8608dacb39eea3f2c495卷已经不见了。

也可通过查看/var/lib/docker/volumes目录证实。

```
root@blockchain:/# cd /var/lib/docker/volumes
root@blockchain:/var/lib/docker/volumes# ls
438c70d81149462db4c35d7e4a67ab14b70d31f566d56abc6cae976f111c1fee  metadata.db
root@blockchain:/var/lib/docker/volumes#
```

从查看到的卷信息可知，当前还有一个438c70d81149462db4c35d7e4a67ab14b70d31f566d56abc6cae976f111c1fee卷存在，正是前面httpd_vol容器预留下来的卷，成为了孤儿卷，因此对于不需要的容器和卷，决定要删除，同时对应的卷也要删除时，在删除容器时带上"-v"参数，一并将无用的卷也删除。下面删除不需要的卷。

⑦删除不需要的卷，采用"docker volume rm 卷名"命令删除不需要的卷。

```
root@blockchain:/# docker volume rm 438c70d81149462db4c35d7e4a67ab14b70d31f566d56abc6cae976f111c1fee
438c70d81149462db4c35d7e4a67ab14b70d31f566d56abc6cae976f111c1fee
```

用docker volume ls命令查看卷信息。

```
root@blockchain:/# docker volume ls
DRIVER                VOLUME NAME
root@blockchain:/#
```

发现卷的列表中已经没有卷了。

还可以批量方式删除不需要的卷，采用如下命令。

```
docker volume rm $(docker volume ls -qf dangling=true)
```

任务总结

容器使用数据卷的基本方法有两种：一个是bind mount；另一个是docker managed volume。二者本质上都是将宿主机的目录映射到容器中。bind mount既可以映射目录也可以是文件；docker managed volume只能映射目录，而且不能提前映射，所映射的目录是在/var/lib/docker/volumes下随机生成的。

bind mount 与 docker managed volume实际上都是使用 host 文件系统的中的某个路径作为mount 源。它们的区别如表3-1所示。

表 3-1 bind mount 与 docker managed volume 的区别

不同点	bind mount	docker managed volume
volume 位置	可任意指定	/var/lib/docker/volumes/...
对已有mount point 影响	隐藏并替换为 volume	原有数据复制到 volume
是否支持单个文件	支持	不支持，只能是目录
权限控制	可设置为只读，默认为读写权限	无控制，均为读写权限
移植性	移植性弱，与 host path 绑定	移植性强，无须指定 host 目录

任务扩展

创建一个目录并设置网页文件，采用bind mount方式挂载到nginx容器中，更改网页内容测试网页的变化。

任务二 通过卷容器挂载数据卷

任务场景

由于业务需求，安安公司已经自己信息系统的业务移植到容器（docker）上面，但是现在需要将宿主机上的数据共享到容器中，有的数据可以长期不发生变化，有的数据需要及时更新，有的数据属于临时挂载使用。假设你是docker工程师，需要完成该工作。

任务描述

本任务学习数据卷的挂载，根据业务的不同，需要学习volume container和data-packed volume container方式数据卷的挂载。

任务目标

◎ 能够描述和实现volume container数据卷挂载
◎ 能够描述和实现data-packed volume container数据卷挂载
◎ 能够总结容器挂载数据卷的方法，描述其主要区别

任务实施

一、通过卷容器挂载卷

卷容器（volume container）是专门为其他容器提供 volume 的容器，提供的卷可以是 bind mount，也可以是 docker managed volume。与 bind mount 相比，不必为每一容器指定 host path，所有 path 都在 volume container 中定义好，容器只需要与 volume container 关联，从而实现容器与 host 卷的挂载。

1. 创建卷容器

用docker create命令创建一个容器，通过bind mount和docker managed volume两种方式挂载卷。

（1）创建一个卷容器

```
root@blockchain:/# docker create --name vc_data -v ~/htdocs:/usr/local/apache2/htdocs -v /other/tools busybox
d56118def0bc1caa34eb95fda78a45be120eb1c94cc4cc7320050814ff957485
```

该命令中 –v ~/htdocs:/usr/local/apache2/htdocs是通过bind mount方式挂载卷，即将宿主机的/htdocs挂载到 vc_data容器的/usr/local/apache2/htdocs目录，–v /other/tools是通过docker managed volume方式挂载，系统会在宿主机/var/lib/docker/volumes文件夹下随机创建一个文件夹挂载到 vc_data容器的 /other/tools文件夹，如果/other/tools文件夹不存在，docker会自动创建该文件夹。

（2）查看新创建的容器进程。

用docker ps –a命令查看vc_data容器信息。

```
root@blockchain:/# docker ps -a
CONTAINER ID    IMAGE       COMMAND     CREATED         STATUS      PORTS     NAMES
d56118def0bc    busybox     "sh"        59 seconds ago  Created               vc_data
```

（3）查看vc_data卷信息

用docker inspect vc_data命令查看vc_data卷挂载信息，这里可加上more参数，显示数据较多，查找Mounts相关信息。

```
root@blockchain:/# docker inspect vc_data
...    （此处省略了部分显示信息）
"Mounts": [
        {
            "Type": "bind",
            "Source": "/root/htdocs",
            "Destination": "/usr/local/apache2/htdocs",
            "Mode": "",
            "RW": true,
            "Propagation": "rprivate"
        },
        {
            "Type": "volume",
            "Name": "786a8cc79ba4926b8743aec08b69ef12def10d72b8eb0a1abfc6758e2511433d",
            "Source": "/var/lib/docker/volumes/786a8cc79ba4926b8743aec08b69ef12def10d72b8eb0a1abfc6758e2511433d/_data",
            "Destination": "/other/tools",
            "Driver": "local",
            "Mode": "",
            "RW": true,
```

```
                    "Propagation": ""
                }
            ],
...（此处省略了部分显示信息）
```

该命令显示信息中有两种类型的卷：一种是bind类型，源路径是/root/htdocs；另一种是volume类型，源路径是/var/lib/docker/volumes/786a8cc79ba4926b8743aec08b69ef12def10d72b8eb0a1abfc6758e2511433d/_data。

可用通过docker volume ls查看创建的volume。

```
root@blockchain:/# docker volume ls
DRIVER    VOLUME NAME
local     786a8cc79ba4926b8743aec08b69ef12def10d72b8eb0a1abfc6758e2511433d
root@blockchain:/#
```

从上面命令可知，创建一个 vc_data容器，状态是Created，通过bind mount将宿主机的~/htdocs目录挂载到容器的/usr/local/apache2/htdocs目录，同时用docker managed volume将宿主机的786a8cc79ba4926b8743aec08b69ef12def10d72b8eb0a1abfc6758e2511433d卷挂载到容器的/other/tools文件夹。

2. 使用已有的卷容器挂载卷

本任务是通过创建一个容器，该容器中挂载好卷，然后其他容器运行时，通过--volumes-from参数挂载该容器已经挂载卷。本操作流程是首先在宿主机上创建好挂载的目录及目录下对应的文件，然后创建卷容器，该卷容器不需要运行，只需要创建即可，接着运行新的容器，命令行中使用--volumes-from参数挂载原来创建的卷容器中的卷，以达到挂载卷的目的。

（1）更改网页文件

```
root@blockchain:/# cd ~/htdocs
root@blockchain:~/htdocs# ls
index.html
root@blockchain:~/htdocs# vi index.html
```

在文件中录入"This is the volume container!"内容并保存。

（2）运行容器使用vc_data容器中的卷

运行3个容器（web1、web2、web3），采用 --volumes-from参数，将vc_data容器中的卷分别挂载到web1、web2、web3容器中。

```
root@blockchain:~/htdocs# docker run -d --name web1 -p 80 --volumes-from vc_data httpd
852a50c114b5f701248c35372e84b28ab4a144235c0cf8b40565cb9938bc342f
root@blockchain:~/htdocs# docker run -d --name web2 -p 80 --volumes-from vc_data httpd
ceb9dc0c323ce0a15ca24137c9737b5cfc49a5d6c8a85b4aaef61b32e7a62ec1
root@blockchain:~/htdocs# docker run -d --name web3 -p 80 --volumes-from vc_data httpd
65a19d49705919c9fb185c3b986bd984875908f8fc58ad88b6bcf69be18dc586
root@blockchain:~/htdocs#
```

(3)查看容器挂载卷的信息

用docker inspect web1命令查看web1容器挂载卷信息。

```
root@blockchain:~/htdocs# docker inspect web1
…（此处省略了部分显示信息）
 "Mounts": [
            {
                "Type": "bind",
                "Source": "/root/htdocs",
                "Destination": "/usr/local/apache2/htdocs",
                "Mode": "",
                "RW": true,
                "Propagation": "rprivate"
            },
            {
                "Type": "volume",
                "Name": "786a8cc79ba4926b8743aec08b69ef12def10d72b8eb0a1abfc6758e2511433d",
                "Source": "/var/lib/docker/volumes/786a8cc79ba4926b8743aec08b69ef12def10d72b8eb0a1abfc6758e2511433d/_data",
                …（此处省略了部分显示信息）
            }
        ],
…（此处省略了部分显示信息）
```

同样可以查看容器web2、web3挂载卷信息,它们是一样的。

(4)查看web1、web2、web3进程

```
root@blockchain:~/htdocs# docker ps
CONTAINER ID        IMAGE              COMMAND             CREATED
STATUS              PORTS              NAMES
  65a19d497059       httpd              "httpd-foreground"   7 minutes ago
Up 7 minutes        0.0.0.0:32771->80/tcp    web3
  ceb9dc0c323c       httpd              "httpd-foreground"   7 minutes ago
Up 7 minutes        0.0.0.0:32770->80/tcp    web2
  852a50c114b5       httpd              "httpd-foreground"   7 minutes ago
Up 7 minutes        0.0.0.0:32769->80/tcp    web1
```

通过查看web1、web2、web3进程,确定3个容器对应的访问端口号。

(5)查看网页文件

分别通过curl命令和网页浏览方式查看网页内容,内容一致,正是卷容器中网页的内容,如图3-8~图3-10所示。

```
root@blockchain:~/htdocs# curl 192.168.47.132:32769
<html><body><h1> This is the volume container!</h1></body></html>
root@blockchain:~/htdocs#
```

图 3-8　web1 页面

```
root@blockchain:~/htdocs# curl 192.168.47.132:32770
<html><body><h1> This is the volume container!</h1></body></html>
```

图 3-9　web2 页面

```
root@blockchain:~/htdocs# curl 192.168.47.132:32771
<html><body><h1> This is the volume container!</h1></body></html>
```

图 3-10　web3 页面

（6）更改卷中网页文件的内容

```
root@blockchain:~/htdocs# vi index.html
```

更改文件内容，录入"This is the content of the page after the volume container has changed！"内容并保存。

（7）再次查看网页内容（见图3-11～图3-13）

```
root@blockchain:~/htdocs# curl 192.168.47.132:32769
<html><body><h1> This is the content of the page after the volume
container has changed!</h1></body></html>
```

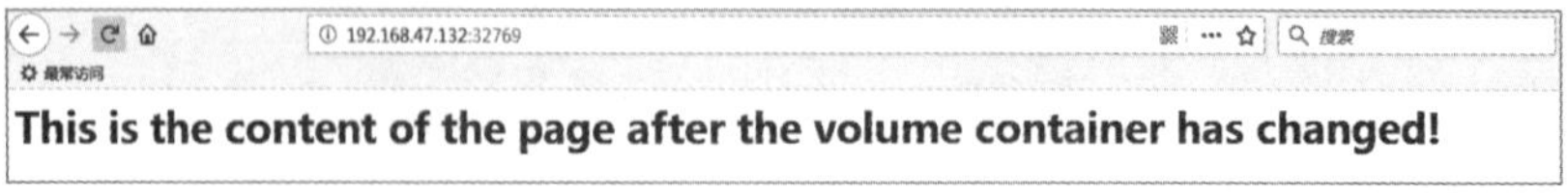

图 3-11　web1 更改后页面

root@blockchain:~/htdocs# curl 192.168.47.132:32770
<html><body><h1> This is the content of the page after the volume container has changed!</h1></body></html>

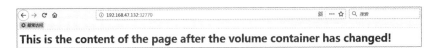

图 3-12　web2 更改后页面

root@blockchain:~/htdocs# curl 192.168.47.132:32771
<html><body><h1> This is the content of the page after the volume container has changed!</h1></body></html>
root@blockchain:~/htdocs#

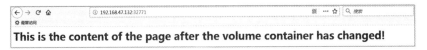

图 3-13　web3 更改后页面

从图3-11～图3-13可以看出，网页内容已经更改为 "This is the content of the page after the volume container has changed!" 信息，通过更改vc_data卷的内容，实现数据卷的共享。这样的好处是任何容器挂载该容器卷中的数据卷，不用再指定具体的path路径，因为所有路径已经在卷容器中定义好了，有利于进行公共资源的共享和管理。

二、通过 data-packed volume container 挂载卷

不论采用bind mount还是docker managed volume的数据卷，归根到底数据还是在 host 里，如果想要将数据完全放到 volume container 中，同时又能与其他容器共享，可以使用 data-packed volume container方式实现。

data-packed volume container 原理是将数据打包到镜像中，然后通过 docker managed volume 共享。data-packed volume container 是自包含的，不依赖 host 提供数据，具有很强的移植性，非常适合只使用静态数据的场景，比如应用的配置信息、Web Server 的静态文件等。下面通过利用dockerfile创建镜像，利用该镜像创建卷容器提供卷的功能，然后创建其他镜像使用该卷容器提供的卷实现数据共享。

1. 创建 dockerfile 文件

root@blockchain:~# vi dockerfile

输入如下内容，然后保存退出。

```
FROM busybox:latest
ADD htdocs /usr/local/apache2/htdocs
VOLUME /usr/local/apache2/htdocs
```

第一行FROM，说明初始镜像是busybox:latest。
第二行ADD，将静态文件添加到容器目录 /usr/local/apache2/htdocs。
第三行VOLUME的作用与 -v 等效，用来创建 docker managed volume，mount point 为 /usr/local/apache2/htdocs，因为这个目录就是 ADD 添加的目录，所以会将已有数据复制到 volume 中。

2. 更改网页文件内容

root@blockchain:~/htdocs# vi index.html

录入"This is the web content provided by the data-packed volume container"内容作为网页显示信息。

3. 利用 dockerfile 生成镜像

使用docker build命令生成dp_container镜像，该镜像用来创建卷容器。

```
root@blockchain:~# docker build -t dp_container .
Sending build context to Docker daemon   47.1kB
Step 1/3 : FROM busybox:latest
 ---> db8ee88ad75f
Step 2/3 : ADD htdocs /usr/local/apache2/htdocs
 ---> e12ed464e941
Step 3/3 : VOLUME /usr/local/apache2/htdocs
 ---> Running in 413888178e9c
Removing intermediate container 413888178e9c
 ---> 0e240fbb08cd
Successfully built 0e240fbb08cd
Successfully tagged dp_container:latest
```

4. 利用 dp_container 镜像创建卷容器

利用dp_container镜像创建一个名称为vc_provid的容器。

```
root@blockchain:~# docker create --name vc_provid dp_container
477fd13bb7e875f8a4fcd63d4578712e7db6d8b7ce85b718b3ca492a195b0f3e
```

查看创建的容器。

```
root@blockchain:~# docker ps -a
CONTAINER ID   IMAGE          COMMAND   CREATED          STATUS     PORTS    NAMES
477fd13bb7e8   dp_container   "sh"      23 minutes ago   Created             vc_provid
```

5. 利用 vc_provid 提供的卷创建容器

利用vc_provid容器提供的卷创建一个名称为vc_web容器，测试网页文件的内容。

```
root@blockchain:~# docker run --name vc_web -d -p 80:80 --volumes-from vc_provid httpd
1ae7e780be3b2891811d066a12e1cec5b16162fa5ff8f1a0485f553cd85d4143
root@blockchain:~#
```

6. 查看网页内容

通过curl 192.168.47.132命令查看网页内容。

```
root@blockchain:~# curl 192.168.47.132
<html><body><h1> This is the web content provided by the data-packed volume container!</h1></body></html>
root@blockchain:~#
```

通过浏览器查看网页内容，如图3-14所示，通过生成的容器卷挂载成功。

项目三 管理容器外加数据卷

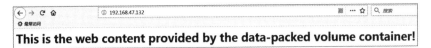

图 3-14 data-packed volume container 挂载测试页面

7. 更改网页内容进行测试

（1）更改网页内容

root@blockchain:~# vi htdocs/index.html

在原来网页内容后面增加"After testing to change the web page"内容，保存退出。

（2）再次查看网页内容

`<html><body><h1> This is the web content provided by the data-packed volume container!</h1></body></html>`

发现更改网页文件的内容并没有及时更新，为什么呢？

原因是容器vc_web容器使用的卷是vc_provid容器提供的，而vc_provid容器使用的是dp_container镜像生成，也就是说，vc_web容器使用的卷是固化在dp_container镜像中的，如果要使用新的网页内容，需要重新由dockerfile生成镜像，利用新生成镜像重新生成容器才能生效。

通过卷容器挂载卷可以认为是bind mount与docker managed volume的扩展，其他容器可以利用卷容器挂载卷的数据，而且可以随时更改卷中的数据，容器的数据也立即更改生效。

data-packed volume container挂载卷的方法是将需要共享的数据卷生成镜像，然后利用该镜像再生成容器，其他容器使用该卷容器挂载卷的数据，此处的操作与卷容器挂载卷一样。与卷容器挂载卷方法不同之处是先创建包含数据源的镜像，另外data-packed volume container挂载卷对生成镜像的数据源进行更改不影响容器的数据卷数据。

创建dockerfile文件，自己定义目录和网页文件，生成镜像运行卷容器，采用data-packed volume container挂载Nginx容器，查看网页内容。

认识和理解 Docker 网络

【项目综述】

根据公司的业务需求，安安公司已经将自己信息系统的业务移植到容器（Docker）上面，随着业务的增加，在维护过程中，需要认识和理解 Docker 网络管理的内容，为了让公司业务顺利运行，需要你对信息部门员工进行 Docker 网络培训，学习 Docker 网络相关知识。

【项目目标】

◎ 能够描述 Docker 网络的类型和应用场景
◎ 能够根据需要配置 Docker 网络
◎ 能够描述 Docker 网络原理和桥接关系
◎ 根据需要能对 Docker 网络进行管理和维护

任务一 认识 Docker 网络

任务场景

由于业务需求，安安公司已将自己信息系统的业务移植到容器（docker）上面，因为维护和公司网络结构设计的需求，需要对信息部门员工进行Docker网络技术的培训，学习Docker网络相关知识，能根据不同场景选择不同的网络类型。

任务描述

本任务学习Docker网络基础知识，Docker网络原理和桥接关系。

任务目标

◎ 能够描述Docker网络的类型和应用场景
◎ 能够根据需求配置Docker网络
◎ 能够描述Docker网络原理和桥接关系
◎ 能够使用命令查看Docker网络属性信息

任务实施

一、查看 Docker 网络类型

Docker 网络的类型分为none、host 和bridge网络，可以采用docker network ls命令查看。

birdge：就如同桥接的switch/hub，使用bridge网络的container会分配一个当前bridge配置的子网IP，可以在通过run创建container时通过 --ip 指定。

host：需要使用 --network=host 参数指定。使用主机网络，此时 container 的网络会附属在主机上，两者是互通的。例如，在container中的服务监听8080端口，则主机的8080端口就会自动映射到这个端口。

none：需要使用 --network=none 参数指定，不分配局域网的IP。

```
root@blockchain:/# docker network ls
NETWORK ID          NAME                DRIVER              SCOPE
fbc837e2c21d        bridge              bridge              local
040ef9dac1b0        host                host                local
2abccc0deebf        none                null                local
root@blockchain:/#
```

二、测试 none 网络

1. 使用 none 网络运行容器

运行一个busybox容器，使用 --network=none参数，查看相关网络信息。

```
root@blockchain:/# docker run -it --network=none busybox
Unable to find image 'busybox:latest' locally
latest: Pulling from library/busybox
Digest: sha256:9f1003c480699be56815db0f8146ad2e22efea85129b5b5983d0e0fb52d9ab70
Status: Downloaded newer image for busybox:latest
/#
```

2. 查看网络配置

```
/# ip l
1: lo: <LOOPBACK,UP,LOWER_UP> mtu 65536 qdisc noqueue qlen 1000
    link/loopback 00:00:00:00:00:00 brd 00:00:00:00:00:00
/# ifconfig
lo        Link encap:Local Loopback
          inet addr:127.0.0.1  Mask:255.0.0.0
          UP LOOPBACK RUNNING  MTU:65536  Metric:1
          RX packets:0 errors:0 dropped:0 overruns:0 frame:0
          TX packets:0 errors:0 dropped:0 overruns:0 carrier:0
          collisions:0 txqueuelen:1000
          RX bytes:0 (0.0 B)  TX bytes:0 (0.0 B)
/#
```

可以看到是一个lo网卡所在的网络，地址是127.0.0.1。

三、查看和测试 docker 的 host 网络

1. 使用 host 网络运行容器

运行一个busybox容器，使用--network=host参数，查看相关网络信息。

```
root@blockchain:/# docker run -it --network=host busybox
```

2. 用 ip l 命令查看当前网卡信息

```
/# ip l
1: lo: <LOOPBACK,UP,LOWER_UP> mtu 65536 qdisc noqueue qlen 1000
    link/loopback 00:00:00:00:00:00 brd 00:00:00:00:00:00
2: ens32: <BROADCAST,MULTICAST,UP,LOWER_UP> mtu 1500 qdisc fq_codel qlen 1000
    link/ether 00:0c:29:58:f5:ad brd ff:ff:ff:ff:ff:ff
3: docker0: <NO-CARRIER,BROADCAST,MULTICAST,UP> mtu 1500 qdisc noqueue
    link/ether 02:42:b8:4b:85:c7 brd ff:ff:ff:ff:ff:ff
/#
```

可以看到当前3个网卡：lo、ens32和docker，这都是主机本身的网卡。

3. 用 ifconfig 命令查看网络配置信息

再用 ifconfig 命令查看当前网卡配置信息，即lo、ens32和docker网卡配置信息。

```
/# ifconfig
docker0   Link encap:Ethernet   HWaddr 02:42:B8:4B:85:C7
          inet addr:172.17.0.1  Bcast:172.17.255.255  Mask:255.255.0.0
          inet6 addr: fe80::42:b8ff:fe4b:85c7/64 Scope:Link
          UP BROADCAST MULTICAST  MTU:1500  Metric:1
          RX packets:206 errors:0 dropped:0 overruns:0 frame:0
          TX packets:241 errors:0 dropped:0 overruns:0 carrier:0
          collisions:0 txqueuelen:0
          RX bytes:782578 (764.2 KiB)  TX bytes:788349 (769.8 KiB)

ens32     Link encap:Ethernet   HWaddr 00:0C:29:58:F5:AD
          inet addr:192.168.47.132  Bcast:192.168.47.255  Mask:255.255.255.0
          inet6 addr: fe80::20c:29ff:fe58:f5ad/64 Scope:Link
          UP BROADCAST RUNNING MULTICAST  MTU:1500  Metric:1
          RX packets:15825 errors:0 dropped:0 overruns:0 frame:0
          TX packets:7493 errors:0 dropped:0 overruns:0 carrier:0
          collisions:0 txqueuelen:1000
          RX bytes:14736222 (14.0 MiB)  TX bytes:827004 (807.6 KiB)

lo        Link encap:Local Loopback
          inet addr:127.0.0.1  Mask:255.0.0.0
          inet6 addr: ::1/128 Scope:Host
          UP LOOPBACK RUNNING  MTU:65536  Metric:1
          RX packets:416 errors:0 dropped:0 overruns:0 frame:0
          TX packets:416 errors:0 dropped:0 overruns:0 carrier:0
          collisions:0 txqueuelen:1000
          RX bytes:40186 (39.2 KiB)  TX bytes:40186 (39.2 KiB)
/#
```

4. 用 ifconfig 查看宿主机的网络信息

再进一步在宿主机上查看与容器的网络信息比较，通过对比，当前busybox容器的网卡配置信息与主机的网卡配置信息一致。

```
root@blockchain:~# ifconfig
docker0: flags=4099<UP,BROADCAST,MULTICAST>  mtu 1500
        inet 172.17.0.1  netmask 255.255.0.0  broadcast 172.17.255.255
        inet6 fe80::42:b8ff:fe4b:85c7  prefixlen 64  scopeid 0x20<link>
        ether 02:42:b8:4b:85:c7  txqueuelen 0  (Ethernet)
        RX packets 206  bytes 782578 (782.5 KB)
        RX errors 0  dropped 0  overruns 0  frame 0
        TX packets 241  bytes 788349 (788.3 KB)
        TX errors 0  dropped 0 overruns 0  carrier 0  collisions 0
ens32: flags=4163<UP,BROADCAST,RUNNING,MULTICAST>  mtu 1500
        inet 192.168.47.132  netmask 255.255.255.0  broadcast 192.168.47.255
        inet6 fe80::20c:29ff:fe58:f5ad  prefixlen 64  scopeid 0x20<link>
        ether 00:0c:29:58:f5:ad  txqueuelen 1000  (Ethernet)
        RX packets 15894  bytes 14743266 (14.7 MB)
        RX errors 0  dropped 0  overruns 0  frame 0
        TX packets 7550  bytes 837557 (837.5 KB)
        TX errors 0  dropped 0 overruns 0  carrier 0  collisions 0
lo: flags=73<UP,LOOPBACK,RUNNING>  mtu 65536
        inet 127.0.0.1  netmask 255.0.0.0
        inet6 ::1  prefixlen 128  scopeid 0x10<host>
        loop  txqueuelen 1000  (Local Loopback)
        RX packets 416  bytes 40186 (40.1 KB)
        RX errors 0  dropped 0  overruns 0  frame 0
        TX packets 416  bytes 40186 (40.1 KB)
        TX errors 0  dropped 0 overruns 0  carrier 0  collisions 0
root@blockchain:~#
```

说明busybox容器与主机使用相同的协议栈。

四、查看和测试 Docker 的 bridge 网络

1. 在宿主机上安装 brctl 命令工具

为了能查看网络桥接关系，安装brctl命令工具，可以方便查看网络的桥接接口状态。

```
root@blockchain:~# apt-get install bridge-utils
```

2. 删除所有容器

为了查看网桥桥接信息，将所有的容器删除。

（1）删除所有退出的容器

```
root@blockchain:~# docker rm -v $(docker ps -aq -f status=exited)
cfbe4272f08c
root@blockchain:~#
```

（2）查看运行的容器

本步骤的目的是删除所有容器，这是为了观察方便，也可以不删除。先查看运行容器，停止掉运行的容器，然后删除。

```
root@blockchain:~# docker ps -a
```

CONTAINER ID	IMAGE	COMMAND	CREATED	STATUS	PORTS	NAMES
5c2467c84c3d	busybox	"sh"	15 minutes ago	Up 15 minutes		funny_clarke

root@blockchain:~#

停止运行的容器。

root@blockchain:~# `docker stop $(docker ps -aq )`
5c2467c84c3d
root@blockchain:~#

再次查看容器运行情况。

root@blockchain:~# `docker ps -a`

CONTAINER ID	IMAGE	COMMAND	CREATED	STATUS	PORTS	NAMES
5c2467c84c3d	busybox	"sh"	18 minutes ago	Exited (137) 41 seconds ago		funny_clarke

root@blockchain:~#

删除退出的容器。

root@blockchain:~# `docker rm -v $(docker ps -aq -f status=exited)`
5c2467c84c3d
root@blockchain:~#

再次查看容器，已经没有容器了。

root@blockchain:~# `docker ps -a`

CONTAINER ID	IMAGE	COMMAND	CREATED	STATUS	PORTS	NAMES

root@blockchain:~#

3. 查看桥接网络

root@blockchain:~# `brctl show`

bridge name	bridge id	STP enabled	interfaces
docker0	8000.0242b84b85c7	no	

root@blockchain:~#

可以看到只有docker0网桥，网桥上没有挂接任何设备。

4. 分析桥接网卡

（1）运行一个新的容器

运行一个新的容器，并指定网络类型是bridge，查看桥接网桥的接口发生变化。

root@blockchain:~# `docker run -it --network=bridge busybox`
/#

在另一个终端中查看短ID为ef78d15aefd7的容器正在运行。

root@blockchain:/# `docker ps`

CONTAINER ID	IMAGE	COMMAND	CREATED	STATUS	PORTS	NAMES
ef78d15aefd7	busybox	"sh"	18 hours ago			

```
Up 18 hours                           serene_mendeleev
root@blockchain:/#
```

(2) 查看桥接信息

在另一个终端中用命令brctl show查看桥接网桥的接口发生变化，发现docker0上有一个veth7561b6d接口，该接口就是busybox容器的网卡。

```
root@blockchain:/# brctl show
bridge name     bridge id              STP enabled      interfaces
docker0         8000.0242b84b85c7      no               veth7561b6d
root@blockchain:/#
```

(3) 查看容器网络接口配置信息

在busybox容器中用ifconfig命令查看容器网络配置信息，重点关注IP地址和MAC地址信息。

```
/# ifconfig
eth0      Link encap:Ethernet   HWaddr 02:42:AC:11:00:02
          inet addr:172.17.0.2  Bcast:172.17.255.255  Mask:255.255.0.0
          UP BROADCAST RUNNING MULTICAST  MTU:1500  Metric:1
          RX packets:12 errors:0 dropped:0 overruns:0 frame:0
          TX packets:0 errors:0 dropped:0 overruns:0 carrier:0
          collisions:0 txqueuelen:0
          RX bytes:976 (976.0 B)  TX bytes:0 (0.0 B)

lo        Link encap:Local Loopback
          inet addr:127.0.0.1  Mask:255.0.0.0
          UP LOOPBACK RUNNING  MTU:65536  Metric:1
          RX packets:0 errors:0 dropped:0 overruns:0 frame:0
          TX packets:0 errors:0 dropped:0 overruns:0 carrier:0
          collisions:0 txqueuelen:1000
          RX bytes:0 (0.0 B)  TX bytes:0 (0.0 B)

/#
```

(4) 查看容器网络接口地址信息

在busybox容器中用ip a命令查看eth0@if19接口，可以看到eth0@if19接口的MAC地址和用ifconfig命令查看的eth0接口的MAC地址是一样的，IP地址也是一样的。

```
/# ip a
1: lo: <LOOPBACK,UP,LOWER_UP> mtu 65536 qdisc noqueue qlen 1000
    link/loopback 00:00:00:00:00:00 brd 00:00:00:00:00:00
    inet 127.0.0.1/8 scope host lo
       valid_lft forever preferred_lft forever
18: eth0@if19: <BROADCAST,MULTICAST,UP,LOWER_UP,M-DOWN> mtu 1500 qdisc noqueue
    link/ether 02:42:ac:11:00:02 brd ff:ff:ff:ff:ff:ff
    inet 172.17.0.2/16 brd 172.17.255.255 scope global eth0
       valid_lft forever preferred_lft forever
/#
```

(5) 查看宿主机的网络配置信息

在宿主机上使用ifconfig命令查看宿主机接口，可以看到有veth7561b6d接口，实际就是busybox网卡在宿主机docker0上的接口，下面进一步验证。

```
root@blockchain:/# ifconfig -a
docker0: flags=4163<UP,BROADCAST,RUNNING,MULTICAST>  mtu 1500
        inet 172.17.0.1  netmask 255.255.0.0  broadcast 172.17.255.255
        inet6 fe80::42:b8ff:fe4b:85c7  prefixlen 64  scopeid 0x20<link>
        ether 02:42:b8:4b:85:c7  txqueuelen 0  (Ethernet)
        RX packets 206  bytes 782578 (782.5 KB)
        RX errors 0  dropped 0  overruns 0  frame 0
        TX packets 243  bytes 788529 (788.5 KB)
        TX errors 0  dropped 0 overruns 0  carrier 0  collisions 0
ens32: flags=4163<UP,BROADCAST,RUNNING,MULTICAST>  mtu 1500
        inet 192.168.47.132  netmask 255.255.255.0  broadcast 192.168.47.255
        inet6 fe80::20c:29ff:fe58:f5ad  prefixlen 64  scopeid 0x20<link>
        ether 00:0c:29:58:f5:ad  txqueuelen 1000  (Ethernet)
        RX packets 16246  bytes 14803613 (14.8 MB)
        RX errors 0  dropped 0  overruns 0  frame 0
        TX packets 7811  bytes 878131 (878.1 KB)
        TX errors 0  dropped 0 overruns 0  carrier 0  collisions 0
lo: flags=73<UP,LOOPBACK,RUNNING>  mtu 65536
        inet 127.0.0.1  netmask 255.0.0.0
        inet6 ::1  prefixlen 128  scopeid 0x10<host>
        loop  txqueuelen 1000  (Local Loopback)
        RX packets 422  bytes 41004 (41.0 KB)
        RX errors 0  dropped 0  overruns 0  frame 0
        TX packets 422  bytes 41004 (41.0 KB)
        TX errors 0  dropped 0 overruns 0  carrier 0  collisions 0
veth7561b6d: flags=4163<UP,BROADCAST,RUNNING,MULTICAST>  mtu 1500
        inet6 fe80::3cd0:2eff:fe8e:5c92  prefixlen 64  scopeid 0x20<link>
        ether 3e:d0:2e:8e:5c:92  txqueuelen 0  (Ethernet)
        RX packets 0  bytes 0 (0.0 B)
        RX errors 0  dropped 0  overruns 0  frame 0
        TX packets 13  bytes 1046 (1.0 KB)
        TX errors 0  dropped 0 overruns 0  carrier 0  collisions 0
```

(6) 在宿主机上查看veth7561b6d接口的对端接口。

```
root@ubuntu-chain0:~# ethtool -S veth7561b6d
NIC statistics:
     peer_ifindex: 18
```

这里命令显示的结果是veth7561b6d接口的对端接口序号为18，与前面在容器中用ip a命令显示的结果对比可知，宿主机接口veth7561b6d和eth0@if19实际上是一对veth pair。veth pair 是一种成对出现的特殊网络设备，可以把它们想象成由一根虚拟网线连接起来的一对网卡，网卡的一头（eth0@if19）在容器中，另一头（veth7561b6d）挂在网桥 docker0 上，其效果就是将 eth0@if19 也挂在了 docker0 上。

(7) 查看宿主机的桥接命名空间

再用命令 docker network inspect bridge 查看信息，可以看到网桥 docker0 的网络配置，网段是 172.17.0.0/16，网关是 172.17.0.1，IP 地址是 172.17.0.2，MAC 地址是 02:42:ac:11:00:02。

```
root@blockchain:/# docker network inspect bridge
[
    {
        "Name": "bridge",
        "Id": "fbc837e2c21da8b3fef2e2cb4564d9d978d5b884ade667581bb0cfbc76e7cf83",
        "Created": "2019-08-05T03:19:32.549186706Z",
        "Scope": "local",
        "Driver": "bridge",
        "EnableIPv6": false,
        "IPAM": {
            "Driver": "default",
            "Options": null,
            "Config": [
                {
                    "Subnet": "172.17.0.0/16",
                    "Gateway": "172.17.0.1"
                }
            ]
        },
        "Internal": false,
        "Attachable": false,
        "Ingress": false,
        "ConfigFrom": {
            "Network": ""
        },
        "ConfigOnly": false,
        "Containers": {
            "ef78d15aefd768630c1c0294899f88d7a01396d4880c19d6a5399488d5dc0ce9": {
                "Name": "serene_mendeleev",
                "EndpointID": "4da01b115da22eb224e26a5a34766f307d132c1c733c5ddc8d9e5a20a76f211f",
                "MacAddress": "02:42:ac:11:00:02",
                "IPv4Address": "172.17.0.2/16",
                "IPv6Address": ""
            }
        },
        "Options": {
```

```
       …     （此处省略了部分显示信息）
       },
       "Labels": {}
    }
]
```

(8) 网络结构图

从以上分析可见，在192.168.47.132主机上，安装Docker时创建了docker0接口，同时创建了docker0网桥，当创建busybox容器时，在docker0网桥上创建了veth7561b6d接口，该接口与busybox容器的eth0@if19接口相连接，形成了虚拟网络连接，eth0@if19接口在busybox容器内部用ifconfig命令查看时表现为eth0接口。容器的桥接结构图如图4-1所示。

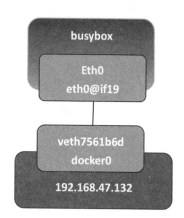

图 4-1　容器的桥接结构

任务总结

①birdge: 就如同桥接的switch/hub，使用bridge网络的container会分配一个当前bridge配置的子网IP，可以在通过run创建container时通过 --ip 指定，默认情况下docker分配的地址是172.17.0.0/16段，网关是172.17.0.1/16。桥接的方法是在宿主机docker0网桥上创建了veth接口，该接口与容器的eth0接口相连接，形成了虚拟网络连接，达到数据通信的功能。

②host: 需要使用 --network=host 参数指定。使用主机网络，此时 container 的网络会附属在主机上，两者是互通的，容器与主机使用相同的协议栈，包括IP地址和mac地址都完全相同。

③none: 需要使用 --network=none 参数指定，不分配局域网的IP，直接使用lo的127.0.0.1地址。

任务扩展

尝试创建多个容器并使用birdge模式，查看容器的网络信息并测试不同容器之间的连通性。

任务二　分析自定义 Docker 网络

任务场景

由于业务需求，安安公司已将自己信息系统的业务移植到容器（Docker）上面，因为维护公司网络的需求，需要自定义Docker网络，创建容器，测试容器之间的连通性，分析Docker网络原理和桥接关系。

任务描述

本任务学习自定义Docker网络，创建容器，测试容器之间的连通性，分析Docker网络原理和桥接关系。

项目四 认识和理解 Docker 网络

任务目标

◎ 能够创建自定义的Docker网络
◎ 能够利用自定义的Docker网络创建容器
◎ 能够分析Docker网络原理和桥接关系
◎ 能够使用命令查看Docker网络属性信息

一、自定义网络

1. 创建一个自己定义的桥接网络

（1）查看宿主机桥接信息

```
root@blockchain:/# brctl show
bridge name     bridge id           STP enabled     interfaces
docker0         8000.0242b84b85c7   no              veth7561b6d
```

从命令上可以看出，当前只有docker0网桥。

（2）创建一个自己的网络

```
root@blockchain:/# docker network create --driver bridge my_net
f3cf41bd2b7d1db3e7d9853a083f2f5e600ab3268857ee87938d77a9e119e0ca
```

命令执行完毕返回的字串就是新创建网络的长ID。

（3）再次查看宿主机桥接信息

```
root@blockchain:/# brctl show
bridge name         bridge id           STP enabled     interfaces
br-f3cf41bd2b7d     8000.0242d2fecba2   no
docker0             8000.0242b84b85c7   no              veth7561b6d
root@blockchain:/#
```

可以看到有一个新的网桥br-f3cf41bd2b7d，可以看到该桥的f3cf41bd2b7d字串就是my_net网桥的短ID。至此，创建了一个新的网桥，该网桥属于bridge类型网桥。

（4）查看网桥的命名空间

采用docker network inspect my_net命令查看my_net网桥具体的配置信息，可以看到该网桥名称、ID、创建时间等信息，分配的网段是172.19.0.0/16，网关是172.19.0.1，暂时该网桥上没有挂接任何容器。

```
root@blockchain:/# docker network inspect my_net
[
    {
        "Name": "my_net",
        "Id": "f3cf41bd2b7d1db3e7d9853a083f2f5e600ab3268857ee87938d77a9e119e0ca",
        "Created": "2019-08-06T03:02:28.044582222Z",
        "Scope": "local",
        "Driver": "bridge",
        "EnableIPv6": false,
        "IPAM": {
```

91

```
                "Driver": "default",
                "Options": {},
                "Config": [
                    {
                        "Subnet": "172.19.0.0/16",
                        "Gateway": "172.19.0.1"
                    }
                ]
        },
        … （此处省略了部分显示信息）
    }
]
root@blockchain:/#
```

2. 创建一个完全自定义的网络

（1）创建my_net29的172.29.0.0/16的网络

```
root@blockchain:/# docker network create --driver bridge --subnet 172.29.0.0/16 --gateway 172.29.0.1 my_net29
b2d3a4288c6ffa69e1a81b2969adee93afe919be9fac23dba2d02ff959053e7f
```

命令执行完毕返回的字串就是新创建网络的长ID。

（2）查看宿主机桥接信息

```
root@blockchain:/# brctl show
bridge name          bridge id              STP enabled      interfaces
br-b2d3a4288c6f      8000.0242c14f74ce      no
br-f3cf41bd2b7d      8000.0242d2fecba2      no
docker0              8000.0242b84b85c7      no               veth7561b6d
```

可以看到有一个新的网桥br-b2d3a4288c6f，可以看到该桥的b2d3a4288c6f字串就是my_net29网桥的短ID。至此，自己指定网段和网关的网桥就创建好了，该网桥属于bridge类型网桥。

（3）查看my_net29网桥的命名空间信息

```
root@blockchain:/# docker network inspect my_net29
[
    {
        "Name": "my_net29",
        "Id": "b2d3a4288c6ffa69e1a81b2969adee93afe919be9fac23dba2d02ff959053e7f",
        "Created": "2019-08-06T03:20:56.295406629Z",
        "Scope": "local",
        "Driver": "bridge",
        "EnableIPv6": false,
        "IPAM": {
            "Driver": "default",
            "Options": {},
            "Config": [
```

```
            {
                "Subnet": "172.29.0.0/16",
                "Gateway": "172.29.0.1"
            }
        ]
    },
    ...（此处省略了部分显示信息）  }
]
```

根据命令输出结果，可以看到该网桥名称、ID、创建时间等信息，分配的网段是172.29.0.0/16，网关是172.29.0.1，暂时该网桥上没有挂接任何容器。

二、使用自定义网络创建容器

1. 使用 my_net 创建一个容器

root@blockchain:/# docker run -it --name=busybox1901 --network=my_net busybox

用ifconfig和ip a命令查看网卡配置信息，可以看到容器busybox1901分配的IP地址是172.19.0.2。

```
/# ifconfig
eth0      Link encap:Ethernet  HWaddr 02:42:AC:13:00:02
          inet addr:172.19.0.2  Bcast:172.19.255.255  Mask:255.255.0.0
          UP BROADCAST RUNNING MULTICAST  MTU:1500  Metric:1
          RX packets:14 errors:0 dropped:0 overruns:0 frame:0
          TX packets:0 errors:0 dropped:0 overruns:0 carrier:0
          collisions:0 txqueuelen:0
          RX bytes:1172 (1.1 KiB)  TX bytes:0 (0.0 B)
lo        Link encap:Local Loopback
          inet addr:127.0.0.1  Mask:255.0.0.0
          UP LOOPBACK RUNNING  MTU:65536  Metric:1
          RX packets:0 errors:0 dropped:0 overruns:0 frame:0
          TX packets:0 errors:0 dropped:0 overruns:0 carrier:0
          collisions:0 txqueuelen:1000
          RX bytes:0 (0.0 B)  TX bytes:0 (0.0 B)
/# ip a
1: lo: <LOOPBACK,UP,LOWER_UP> mtu 65536 qdisc noqueue qlen 1000
    link/loopback 00:00:00:00:00:00 brd 00:00:00:00:00:00
    inet 127.0.0.1/8 scope host lo
       valid_lft forever preferred_lft forever
23: eth0@if24: <BROADCAST,MULTICAST,UP,LOWER_UP,M-DOWN> mtu 1500 qdisc noqueue
    link/ether 02:42:ac:13:00:02 brd ff:ff:ff:ff:ff:ff
    inet 172.19.0.2/16 brd 172.19.255.255 scope global eth0
       valid_lft forever preferred_lft forever
/#
```

下面测试与互联网的通信，可见一切正常。

```
/# ping www.baidu.com
PING www.baidu.com (14.215.177.39): 56 data bytes
64 bytes from 14.215.177.39: seq=0 ttl=127 time=8.222 ms
64 bytes from 14.215.177.39: seq=2 ttl=127 time=8.544 ms
^C
--- www.baidu.com ping statistics ---
2 packets transmitted, 2 packets received, 0% packet loss
round-trip min/avg/max = 8.222/8.502/8.740 ms
/#
```

至此，使用my_net网络创建了busybox1901容器，通过测试能与外网正常通信。需要注意的是，在创建容器时需要加上--network参数，本例中还用了--name参数，用来指定容器的名称。

2. 使用 my_net29 创建一个容器

下面利用my_net29创建容器busybox2901，观察IP地址和与外网的连通性。

```
root@blockchain:/# docker run -it --name=busybox2901 --network=my_net29 busybox
/# ifconfig
eth0      Link encap:Ethernet  HWaddr 02:42:AC:1D:00:02
          inet addr:172.29.0.2  Bcast:172.29.255.255  Mask:255.255.0.0
          UP BROADCAST RUNNING MULTICAST  MTU:1500  Metric:1
          RX packets:14 errors:0 dropped:0 overruns:0 frame:0
          TX packets:0 errors:0 dropped:0 overruns:0 carrier:0
          collisions:0 txqueuelen:0
          RX bytes:1172 (1.1 KiB)  TX bytes:0 (0.0 B)

lo        Link encap:Local Loopback
          inet addr:127.0.0.1  Mask:255.0.0.0
          UP LOOPBACK RUNNING  MTU:65536  Metric:1
          RX packets:0 errors:0 dropped:0 overruns:0 frame:0
          TX packets:0 errors:0 dropped:0 overruns:0 carrier:0
          collisions:0 txqueuelen:1000
          RX bytes:0 (0.0 B)  TX bytes:0 (0.0 B)

/# ip a
1: lo: <LOOPBACK,UP,LOWER_UP> mtu 65536 qdisc noqueue qlen 1000
    link/loopback 00:00:00:00:00:00 brd 00:00:00:00:00:00
    inet 127.0.0.1/8 scope host lo
       valid_lft forever preferred_lft forever
25: eth0@if26: <BROADCAST,MULTICAST,UP,LOWER_UP,M-DOWN> mtu 1500 qdisc noqueue
    link/ether 02:42:ac:1d:00:02 brd ff:ff:ff:ff:ff:ff
    inet 172.29.0.2/16 brd 172.29.255.255 scope global eth0
       valid_lft forever preferred_lft forever
/# ping www.baidu.com
PING www.baidu.com (14.215.177.38): 56 data bytes
```

```
64 bytes from 14.215.177.38: seq=0 ttl=127 time=7.073 ms
64 bytes from 14.215.177.38: seq=1 ttl=127 time=6.916 ms
^C
--- www.baidu.com ping statistics ---
2 packets transmitted, 2 packets received, 0% packet loss
round-trip min/avg/max = 6.916/6.984/7.073 ms
/#
```

3. 手动直接指定容器 IP 地址创建容器

```
root@blockchain:/# docker run -it --name=busybox2902 --network=my_net29 --ip 172.29.0.12 busybox
/# ifconfig
eth0      Link encap:Ethernet  HWaddr 02:42:AC:1D:00:0C
          inet addr:172.29.0.12  Bcast:172.29.255.255  Mask:255.255.0.0
          UP BROADCAST RUNNING MULTICAST  MTU:1500  Metric:1
          RX packets:10 errors:0 dropped:0 overruns:0 frame:0
          TX packets:0 errors:0 dropped:0 overruns:0 carrier:0
          collisions:0 txqueuelen:0
          RX bytes:796 (796.0 B)  TX bytes:0 (0.0 B)
lo        Link encap:Local Loopback
          inet addr:127.0.0.1  Mask:255.0.0.0
          UP LOOPBACK RUNNING  MTU:65536  Metric:1
          RX packets:0 errors:0 dropped:0 overruns:0 frame:0
          TX packets:0 errors:0 dropped:0 overruns:0 carrier:0
          collisions:0 txqueuelen:1000
          RX bytes:0 (0.0 B)  TX bytes:0 (0.0 B)
/# ip a
1: lo: <LOOPBACK,UP,LOWER_UP> mtu 65536 qdisc noqueue qlen 1000
    link/loopback 00:00:00:00:00:00 brd 00:00:00:00:00:00
    inet 127.0.0.1/8 scope host lo
       valid_lft forever preferred_lft forever
29: eth0@if30: <BROADCAST,MULTICAST,UP,LOWER_UP,M-DOWN> mtu 1500 qdisc noqueue
    link/ether 02:42:ac:1d:00:0c brd ff:ff:ff:ff:ff:ff
    inet 172.29.0.12/16 brd 172.29.255.255 scope global eth0
       valid_lft forever preferred_lft forever
/#
```

4. 分析网络结构

使用 brctl show 命令查看当前桥接情况。

```
root@blockchain:/# brctl show
bridge name         bridge id              STP enabled      interfaces
br-b2d3a4288c6f     8000.0242c14f74ce      no               veth364e838
                                                            veth86b12cb
br-f3cf41bd2b7d     8000.0242d2fecba2      no               veth9e05f29
```

| docker0 | 8000.0242b84b85c7 | no | veth7561b6d |

root@blockchain:/#

用docker network inspect my_net29命令查看my_net29信息，可以看到busybox2901容器和busybox2902容器的IP地址信息。

```
root@blockchain:/# docker network inspect my_net29
[
    {
        "Name": "my_net29",
        "Id": "b2d3a4288c6ffa69e1a81b2969adee93afe919be9fac23dba2d02ff959053e7f",
        "Created": "2019-08-06T03:20:56.295406629Z",
        "Scope": "local",
        "Driver": "bridge",
        "EnableIPv6": false,
        "IPAM": {
            "Driver": "default",
            "Options": {},
            "Config": [
                {
                    "Subnet": "172.29.0.0/16",
                    "Gateway": "172.29.0.1"
                }
            ]
        },
        "Internal": false,
        "Attachable": false,
        "Ingress": false,
        "ConfigFrom": {
            "Network": ""
        },
        "ConfigOnly": false,
        "Containers": {
            "983355ace4c15b5987b1e6a5ec84667d5a9110d9d3c9f198b654d342c38937dc": {
                "Name": "busybox2901",
                "EndpointID": "51f9b7638d764e45dedad4eb60c1473dd418bcb2788f2fa0d410902820b92773",
                "MacAddress": "02:42:ac:1d:00:02",
                "IPv4Address": "172.29.0.2/16",
                "IPv6Address": ""
            },
            "a1ac543819d19d705ccf3d8df3a5a7e9d99b293995f3b309af9e6d5222faaf11": {
                "Name": "busybox2902",
```

```
                "EndpointID": "72259055a1d0b8b115e5b880dbbb0843cfa5e7c8e7
3967e1713c8b2f08f40df6",
                "MacAddress": "02:42:ac:1d:00:0c",
                "IPv4Address": "172.29.0.12/16",
                "IPv6Address": ""
            }
        },
        "Options": {},
        "Labels": {}
    }
]
root@blockchain:/#
```

从以上分析可知,br-b2d3a4288c6f桥(my_net29网络)上有veth86b12cb和veth364e838接口,br-f3cf41bd2b7d桥(my_net网络)上有veth9e05f29接口,从ifconfig命令和ip a命令查看到的信息可知,br-b2d3a4288c6f桥上的veth86b12cb接口连接的是busybox2901容器,veth364e838接口连接的是busybox2902容器,br-f3cf41bd2b7d桥上的veth9e05f29接口连接的是busybox1901容器。图4-2所示为网络拓扑结构图桥接关系图。

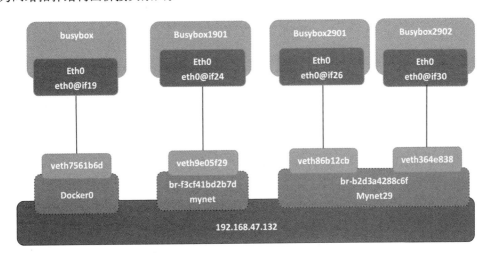

图 4-2 网络拓扑结构图桥接关系图

三、测试容器之间的通信

1. 在 busybox2901 容器上测试到 busybox2902 的连通性

```
root@blockchain:/# docker exec -it  busybox2901  /bin/sh
/# ifconfig eth0
eth0      Link encap:Ethernet   HWaddr 02:42:AC:1D:00:02
          inet addr:172.29.0.2  Bcast:172.29.255.255  Mask:255.255.0.0
          UP BROADCAST RUNNING MULTICAST  MTU:1500  Metric:1
          RX packets:278 errors:0 dropped:0 overruns:0 frame:0
          TX packets:257 errors:0 dropped:0 overruns:0 carrier:0
          collisions:0 txqueuelen:0
          RX bytes:25905 (25.2 KiB)  TX bytes:24184 (23.6 KiB)
```

```
/# ping -c 3 172.29.0.12
PING 172.29.0.12 (172.29.0.12): 56 data bytes
64 bytes from 172.29.0.12: seq=0 ttl=64 time=0.327 ms
64 bytes from 172.29.0.12: seq=1 ttl=64 time=0.201 ms
--- 172.29.0.12 ping statistics ---
2 packets transmitted, 2 packets received, 0% packet loss
round-trip min/avg/max = 0.181/0.236/0.327 ms
/#
```

从测试的结果上可以看到busybox2901和busybox2902能正常通信，原因是两个容器属于同一个网络。

2. 启动 busybox1901 容器测试到 busybox2901 和 busybox2902 的连通性

（1）查看容器运行状况

```
root@blockchain:/# docker ps
CONTAINER ID        IMAGE              COMMAND             CREATED              STATUS              PORTS               NAMES
a1ac543819d1        busybox            "sh"                17 hours ago         Up 17 hours                             busybox2902
983355ace4c1        busybox            "sh"                18 hours ago         Up 18 hours                             busybox2901
e77573e0bc86        busybox            "sh"                19 hours ago         Up 19 hours                             busybox1901
ef78d15aefd7        busybox            "sh"                40 hours ago         Up 40 hours                             serene_mendeleev
root@blockchain:/#
```

从列表中可以看到busybox2901和busybox2902正常运行。

（2）启动busybox1901并测试到busybox2901和busybox2902的连通性

```
root@blockchain:/# docker restart busybox1901
busybox1901
root@blockchain:/# docker exec -it busybox1901 /bin/sh
/# ifconfig eth0
eth0      Link encap:Ethernet  HWaddr 02:42:AC:13:00:02
          inet addr:172.19.0.2  Bcast:172.19.255.255  Mask:255.255.0.0
          UP BROADCAST RUNNING MULTICAST  MTU:1500  Metric:1
          RX packets:41 errors:0 dropped:0 overruns:0 frame:0
          TX packets:7 errors:0 dropped:0 overruns:0 carrier:0
          collisions:0 txqueuelen:0
          RX bytes:3182 (3.1 KiB)  TX bytes:524 (524.0 B)
/# ping -c 3 172.29.0.2
PING 172.29.0.2 (172.29.0.2): 56 data bytes
…  （此处省略了部分显示信息）
 172.29.0.2 ping statistics ---
3 packets transmitted, 0 packets received, 100% packet loss
/# ping -c 3 172.29.0.12
```

```
PING 172.29.0.12 (172.29.0.12): 56 data bytes
…  （此处省略了部分显示信息）
172.29.0.12 ping statistics ---
3 packets transmitted, 0 packets received, 100% packet loss
/#
```

从测试的结果上可以看到busybox1901到busybox2901和busybox2902不通，也就是busybox1901在my_net上，busybox2901和busybox2902在my_net29网络上，不同网络上的容器是不能直接通信的。下面继续研究不同网络上的容器通信问题。

3. 测试不同网络上容器间的通信

运用增加容器网卡实现不同网络上的容器通信：在busybox1901容器上增加一块网卡，连接到my_net29网络上，测试busybox1901容器到busybox2901和busybox2902容器的连通性。

① 查看当前容器进程。

```
root@blockchain:/# docker ps
CONTAINER ID        IMAGE              COMMAND             CREATED             STATUS              PORTS               NAMES
   a1ac543819d1     busybox            "sh"                17 hours ago        Up 17 hours                             busybox2902
   983355ace4c1     busybox            "sh"                18 hours ago        Up 18 hours                             busybox2901
   e77573e0bc86     busybox            "sh"                19 hours ago        Up 19 hours                             busybox1901
   ef78d15aefd7     busybox            "sh"                40 hours ago        Up 40 hours                             serene_mendeleev
```

② 将busybox1901容器增加一块网卡，连接到my_net29网络。

```
root@blockchain:/# docker network connect my_net29  busybox1901
```

③ 进入到busybox1901容器中。

```
root@blockchain:/# docker exec -it  busybox1901 /bin/sh
```

④ 用ifconfig命令显示busybox1901容器的网络配置信息。

```
/# ifconfig
eth0      Link encap:Ethernet   HWaddr 02:42:AC:13:00:02
          inet addr:172.19.0.2  Bcast:172.19.255.255  Mask:255.255.0.0
          UP BROADCAST RUNNING MULTICAST  MTU:1500  Metric:1
          RX packets:46 errors:0 dropped:0 overruns:0 frame:0
          TX packets:14 errors:0 dropped:0 overruns:0 carrier:0
          collisions:0 txqueuelen:0
          RX bytes:3504 (3.4 KiB)  TX bytes:1154 (1.1 KiB)
eth1      Link encap:Ethernet   HWaddr 02:42:AC:1D:00:03
          inet addr:172.29.0.3  Bcast:172.29.255.255  Mask:255.255.0.0
          UP BROADCAST RUNNING MULTICAST  MTU:1500  Metric:1
          RX packets:12 errors:0 dropped:0 overruns:0 frame:0
          TX packets:0 errors:0 dropped:0 overruns:0 carrier:0
```

```
            collisions:0 txqueuelen:0
            RX bytes:936 (936.0 B)  TX bytes:0 (0.0 B)
lo          Link encap:Local Loopback
            inet addr:127.0.0.1  Mask:255.0.0.0
            UP LOOPBACK RUNNING  MTU:65536  Metric:1
            RX packets:4 errors:0 dropped:0 overruns:0 frame:0
            TX packets:4 errors:0 dropped:0 overruns:0 carrier:0
            collisions:0 txqueuelen:1000
            RX bytes:322 (322.0 B)  TX bytes:322 (322.0 B)
/#
```

这里多了一个eth1网卡，IP地址是172.29.0.3，正是刚才用docker network connect my_net29 busybox1901命令新增加的网卡，并分配了IP地址172.29.0.3。

⑤ 用ip a命令显示busybox1901容器的网络配置信息。

```
/# ip a
1: lo: <LOOPBACK,UP,LOWER_UP> mtu 65536 qdisc noqueue qlen 1000
    link/loopback 00:00:00:00:00:00 brd 00:00:00:00:00:00
    inet 127.0.0.1/8 scope host lo
       valid_lft forever preferred_lft forever
23: eth0@if24: <BROADCAST,MULTICAST,UP,LOWER_UP,M-DOWN> mtu 1500 qdisc noqueue
    link/ether 02:42:ac:13:00:02 brd ff:ff:ff:ff:ff:ff
    inet 172.19.0.2/16 brd 172.19.255.255 scope global eth0
       valid_lft forever preferred_lft forever
33: eth1@if34: <BROADCAST,MULTICAST,UP,LOWER_UP,M-DOWN> mtu 1500 qdisc noqueue
    link/ether 02:42:ac:1d:00:03 brd ff:ff:ff:ff:ff:ff
    inet 172.29.0.3/16 brd 172.29.255.255 scope global eth1
       valid_lft forever preferred_lft forever
/#
```

从接口上看eth1@if34接口就是新增加的接口，IP地址是172.29.0.3。

⑥ 测试busybox1901到busybox2901和busybox2902的连通性。进入busybox2901容器并查看该容器的IP地址。

```
root@blockchain:/#  docker exec -it  busybox2901 /bin/sh
/# ifconfig
eth0    Link encap:Ethernet  HWaddr 02:42:AC:1D:00:02
        inet addr:172.29.0.2  Bcast:172.29.255.255  Mask:255.255.0.0
        UP BROADCAST RUNNING MULTICAST  MTU:1500  Metric:1
        RX packets:291 errors:0 dropped:0 overruns:0 frame:0
        TX packets:273 errors:0 dropped:0 overruns:0 carrier:0
        collisions:0 txqueuelen:0
        RX bytes:26871 (26.2 KiB)  TX bytes:25360 (24.7 KiB)
lo      Link encap:Local Loopback
        inet addr:127.0.0.1  Mask:255.0.0.0
        UP LOOPBACK RUNNING  MTU:65536  Metric:1
```

```
            RX packets:4 errors:0 dropped:0 overruns:0 frame:0
            TX packets:4 errors:0 dropped:0 overruns:0 carrier:0
            collisions:0 txqueuelen:1000
            RX bytes:379 (379.0 B)  TX bytes:379 (379.0 B)
/#
```

进入busybox2902容器并查看该容器的IP地址。

```
root@blockchain:/# docker exec -it busybox2902 /bin/sh
/# ifconfig
eth0        Link encap:Ethernet  HWaddr 02:42:AC:1D:00:0C
            inet addr:172.29.0.12  Bcast:172.29.255.255  Mask:255.255.0.0
            UP BROADCAST RUNNING MULTICAST  MTU:1500  Metric:1
            RX packets:24 errors:0 dropped:0 overruns:0 frame:0
            TX packets:5 errors:0 dropped:0 overruns:0 carrier:0
            collisions:0 txqueuelen:0
            RX bytes:1776 (1.7 KiB)  TX bytes:378 (378.0 B)
lo          Link encap:Local Loopback
            inet addr:127.0.0.1  Mask:255.0.0.0
            UP LOOPBACK RUNNING  MTU:65536  Metric:1
            RX packets:0 errors:0 dropped:0 overruns:0 frame:0
            TX packets:0 errors:0 dropped:0 overruns:0 carrier:0
            collisions:0 txqueuelen:1000
            RX bytes:0 (0.0 B)  TX bytes:0 (0.0 B)
/#
```

进入busybox1901容器并查看该容器的IP地址,并测试到busybox2901和busybox2902的连通性。

```
root@blockchain:/# docker exec -it busybox1901 /bin/sh
/# ifconfig
eth0        Link encap:Ethernet  HWaddr 02:42:AC:13:00:02
            inet addr:172.19.0.2  Bcast:172.19.255.255  Mask:255.255.0.0
            UP BROADCAST RUNNING MULTICAST  MTU:1500  Metric:1
            RX packets:46 errors:0 dropped:0 overruns:0 frame:0
            TX packets:14 errors:0 dropped:0 overruns:0 carrier:0
            collisions:0 txqueuelen:0
            RX bytes:3504 (3.4 KiB)  TX bytes:1154 (1.1 KiB)
eth1        Link encap:Ethernet  HWaddr 02:42:AC:1D:00:03
            inet addr:172.29.0.3  Bcast:172.29.255.255  Mask:255.255.0.0
            UP BROADCAST RUNNING MULTICAST  MTU:1500  Metric:1
            RX packets:12 errors:0 dropped:0 overruns:0 frame:0
            TX packets:0 errors:0 dropped:0 overruns:0 carrier:0
            collisions:0 txqueuelen:0
            RX bytes:936 (936.0 B)  TX bytes:0 (0.0 B)
lo          Link encap:Local Loopback
            inet addr:127.0.0.1  Mask:255.0.0.0
```

```
                UP LOOPBACK RUNNING  MTU:65536  Metric:1
                RX packets:4 errors:0 dropped:0 overruns:0 frame:0
                TX packets:4 errors:0 dropped:0 overruns:0 carrier:0
                collisions:0 txqueuelen:1000
                RX bytes:322 (322.0 B)  TX bytes:322 (322.0 B)
/# ping -c 3 172.29.0.2
PING 172.29.0.2 (172.29.0.2): 56 data bytes
64 bytes from 172.29.0.2: seq=0 ttl=64 time=0.780 ms
64 bytes from 172.29.0.2: seq=1 ttl=64 time=0.194 ms
64 bytes from 172.29.0.2: seq=2 ttl=64 time=0.185 ms
--- 172.29.0.2 ping statistics ---
3 packets transmitted, 3 packets received, 0% packet loss
round-trip min/avg/max = 0.185/0.386/0.780 ms
/# ping -c 3 172.29.0.12
PING 172.29.0.12 (172.29.0.12): 56 data bytes
64 bytes from 172.29.0.12: seq=0 ttl=64 time=0.510 ms
64 bytes from 172.29.0.12: seq=1 ttl=64 time=0.227 ms
64 bytes from 172.29.0.12: seq=2 ttl=64 time=0.094 ms
--- 172.29.0.12 ping statistics ---
3 packets transmitted, 3 packets received, 0% packet loss
round-trip min/avg/max = 0.094/0.277/0.510 ms
/#
```

从测试结果上可以看到，在busybox1901容器上增加一块网卡，连接到my_net29网络上，busybox1901容器到busybox2901和busybox2902容器是可以相互通信的。

4. 分析容器与外部网络之间的通信

（1）用容器ping外部网络（互联网或者组织内的其他主机）

用busybox1901容器测试到www.baidu.com的连通性，可以看到可以正常通信。

```
/# ping -c 3 www.baidu.com
PING www.baidu.com (14.215.177.39): 56 data bytes
64 bytes from 14.215.177.39: seq=0 ttl=127 time=9.758 ms
64 bytes from 14.215.177.39: seq=1 ttl=127 time=8.691 ms
64 bytes from 14.215.177.39: seq=2 ttl=127 time=6.772 ms
--- www.baidu.com ping statistics ---
3 packets transmitted, 3 packets received, 0% packet loss
round-trip min/avg/max = 6.772/8.407/9.758 ms
```

（2）分析该连通性使用的技术

① 查看nat转换表，在宿主机上查看nat转换表。

```
root@ubuntu-chain0:~# iptables -t nat -S
-P PREROUTING ACCEPT
-P INPUT ACCEPT
-P OUTPUT ACCEPT
-P POSTROUTING ACCEPT
-N DOCKER
```

```
-A PREROUTING -m addrtype --dst-type LOCAL -j DOCKER
-A OUTPUT ! -d 127.0.0.0/8 -m addrtype --dst-type LOCAL -j DOCKER
-A POSTROUTING -s 172.17.0.0/16 ! -o docker0 -j MASQUERADE
-A POSTROUTING -s 172.18.0.0/16 ! -o docker_gwbridge -j MASQUERADE
-A POSTROUTING -s 172.19.0.0/16 ! -o br-f3cf41bd2b7d -j MASQUERADE
-A DOCKER -i docker0 -j RETURN
-A DOCKER -i docker_gwbridge -j RETURN
-A DOCKER -i br-dd34a4c9d975 -j RETURN
-A DOCKER -i br-234927a2090c -j RETURN
```

② 查看路由信息。

```
root@ubuntu-chain0:~# ip r
default via 192.168.47.2 dev ens32 proto dhcp src 192.168.47.132 metric 100
172.17.0.0/16 dev docker0 proto kernel scope link src 172.17.0.1 linkdown
172.18.0.0/16 dev docker_gwbridge proto kernel scope link src 172.18.0.1 linkdown
172.19.0.0/16 dev br-f3cf41bd2b7d proto kernel scope link src 172.19.0.1
192.168.47.0/24 dev ens32 proto kernel scope link src 192.168.47.137
192.168.47.2 dev ens32 proto dhcp scope link src 192.168.47.137 metric 100
```

③ 查看br-f3cf41bd2b7d上面的ping包。

```
root@ubuntu-chain0:~# tcpdump -i br-f3cf41bd2b7d -n icmp
tcpdump: verbose output suppressed, use -v or -vv for full protocol decode
listening on br-dd34a4c9d975, link-type EN10MB (Ethernet), capture size 262144 bytes
 08:36:44.010599 IP 172.19.0.2 > 14.215.177.39: ICMP echo request, id 7936, seq 0, length 64
 08:36:44.017299 IP 14.215.177.39 > 172.19.0.2: ICMP echo reply, id 7936, seq 0, length 64
 08:36:45.011512 IP 172.19.0.2 > 14.215.177.39: ICMP echo request, id 7936, seq 1, length 64
 08:36:45.017882 IP 14.215.177.39 > 172.19.0.2: ICMP echo reply, id 7936, seq 1, length 64
 08:36:46.015671 IP 172.19.0.2 > 14.215.177.39: ICMP echo request, id 7936, seq 2, length 64
 08:36:46.025291 IP 14.215.177.39 > 172.19.0.2: ICMP echo reply, id 7936, seq 2, length 64
```

④ 查看ens32上面的ping包。

```
root@ubuntu-chain0:~# tcpdump -i ens32 -n icmp
tcpdump: verbose output suppressed, use -v or -vv for full protocol decode
listening on ens32, link-type EN10MB (Ethernet), capture size 262144 bytes
 08:38:59.713370 IP 192.168.47.132 > 14.215.177.39: ICMP echo request, id 8192, seq 0, length 64
 08:38:59.720854 IP 14.215.177.39 > 192.168.47.137: ICMP echo reply, id
```

```
8192, seq 0, length 64
    08:39:00.713885 IP 192.168.47.132 > 14.215.177.39: ICMP echo request, id
8192, seq 1, length 64
    08:39:00.720172 IP 14.215.177.39 > 192.168.47.137: ICMP echo reply, id
8192, seq 1, length 64
    08:39:01.714929 IP 192.168.47.132> 14.215.177.39: ICMP echo request, id
8192, seq 2, length 64
    08:39:01.721109 IP 14.215.177.39 > 192.168.47.132: ICMP echo reply, id
8192, seq 2, length 64
```

从以上查看可以得到一个结论，容器的地址通过地址转换的方式访问外部网络，容器的地址转换成宿主机的IP地址实现的。

例如本任务中，容器的IP地址（172.19.0.2）访问百度（14.215.177.39），然后网桥br-f3cf41bd2b7d发现访问的目的地址是外部网络，就交给iptables（NAT处理），iptables通过NAT地址转换技术将宿主机的ens32地址（192.168.47.132）替换掉容器的IP地址172.19.0.2访问百度，百度的应答送给宿主机的ens32地址（192.168.47.132），再通过地址转换，将目的地址宿主机的ens32地址（192.168.47.132）替换成容器的IP地址172.19.0.2，由br-f3cf41bd2b7d转发给容器，实现了容器与外部网络的通信。

任务总结

① 不管是使用默认的docker网桥还是使用自定义的docker网络创建容器，网桥都是通过一对veth pair与容器相连的。

② 同一个网络上的容器是可以相互通信的，不同网络上的容器之间是不可以相互通信的，要想实现不同网络上的容器之间的相互通信，需要在容器上添加网卡连接到要通信的网络。

③ 容器和外部网络之间的通信是通过NAT地址转换完成的，容器的IP地址会转换成宿主机的IP地址与外部网络进行通信。

④ IP命令由iproute2套件提供，功能强大，有很多子命令，本任务中用到了a和r子命令。ip命令和ifconfig类似，旨在取代ifconfig。ifconfig是net-tools中已被废弃使用的一个命令，已经不再维护。

⑤ brctl 命令由bridge-utile安装包提供，用来管理以太网桥，在内核中建立、维护、检查网桥配置，有很多子命令。

⑥ docker network inspect是查看网络命名空间命令，可以查看详细的Docker网络信息。

⑦ ethtool是用于查询及设置网卡参数的命令，本任务使用了-s参数，显示统计信息和对端的接口的序号。

任务扩展

本任务中同一个网络上的容器是可以相互通信的，不同网络上的容器之间是不可以相互通信的，所有容器均在同一个宿主机中。思考：如果在相互连接能正常通信的不同宿主机上创建相同的网络，位于这些网络上的容器之间能相互通信吗？

使用 Compose 编排服务

【项目综述】

根据公司的业务需求，安安公司已经将自己信息系统的业务移植到容器上面。由于业务发展，需要在宿主机上运行多个容器实现服务功能，由于采用前期单一运行容器的方式比较麻烦，效率低，现需要技术人员采用 docker-compose 编排技术，实现快速部署服务。

【项目目标】

◎ 能够根据业务需要设计编排容器
◎ 能够根据业务需求编写 Compose 配置文件
◎ 能够编排和测试运行容器

任务一　使用 Compose 实现高可用 Web 网站建设

任务场景

由于业务需求，安安公司需要部署高可用的Web服务，需要对Web服务进行冗余设计，当一个容器意外终止时，不影响业务的持续，现需要技术人员采用docker-compose技术编排容器，实现高可用Web网站建设。

任务描述

本任务学习采用docker-compose编排技术，实现高可用Web网站建设，前端采用haproxy负载均衡器实现Web服务，后端用两个Nginx容器实例提供Web服务。

任务目标

◎ 能够根据高可用的Web服务需求，设计高可用架构
◎ 能够识别docker-compose配置文件命令并进行释义
◎ 了解和知道docker-compose配置文件的格式
◎ 能够根据需求编写docker-compose配置文件
◎ 能够使用命令编排和测试高可用的Web服务

任务实施

一、认识 Compose

Docker-Compose项目是Docker官方的开源项目，负责实现对Docker容器集群的快速编排。

Docker-Compose将所管理容器分为3个层次，分别是工程（project）、服务（service）及容器（container）。Docker-Compose运行目录下的所有文件（docker-compose.yml，extends文件或环境变量文件等）组成一个工程，若无特殊指定工程名即为当前目录名。一个工程当中可包含多个服务，每个服务中定义了容器运行的镜像、参数、依赖。一个服务当中可包括多个容器实例，Docker-Compose并没有解决负载均衡的问题，因此需要借助其他工具实现服务发现及负载均衡。

Docker-Compose的工程配置文件默认为docker-compose.yml，可通过环境变量COMPOSE_FILE或-f参数自定义配置文件，其定义了多个有依赖关系的服务及每个服务运行的容器。

使用一个Dockerfile模板文件，可以让用户很方便地定义一个单独的应用容器。在工作中，经常会碰到需要多个容器相互配合来完成某项任务的情况。例如，要实现一个Web项目，除了Web服务容器本身，往往还需要加上后端的数据库服务容器，甚至还包括负载均衡容器等。

Docker-Compose允许用户通过一个单独的docker-compose.yml模板文件（YAML 格式）来定义一组相关联的应用容器为一个项目（project）。

Docker-Compose项目由Python编写，调用Docker服务提供的API来对容器进行管理。因此，只要所操作的平台支持Docker API，就可以在其上利用Docker-Compose来进行编排管理。

Docker-Compose 是 Docker 容器进行编排的工具，定义和运行多容器的应用，可以一条命令启动多个容器，使用Docker Docker-Compose不再需要使用shell脚本来启动容器。

Docker-Compose 通过一个配置文件来管理多个Docker容器，在配置文件中，所有的容器通过services来定义，然后使用docker-compose脚本来启动、停止和重启应用和应用中的服务及所有依赖服务的容器，非常适合组合使用多个容器进行开发的场景。

使用Docker-Compose需要了解和知道yaml文件格式，docker-compose工具工作的时候需要使用一个配置文件，默认的名字为docker-compose.yaml；知道docker-compose中常用关键字及docker-compose操作命令。

二、网络架构设计

根据公司的Web网站建设需求，采用图5-1所示，架构设计，前端通过haproxy负载均衡器实现Web服务，后端用两个Nginx容器实例提供Web服务。

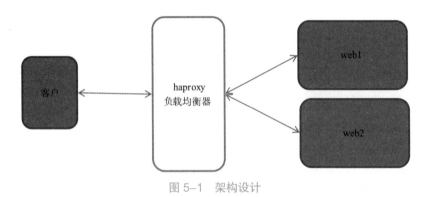

图 5-1　架构设计

三、安装 Docker-Compose 工具包

1. 安装环境查看

```
root@blockchain:/usr/local/bin# uname -a
Linux blockchain 4.15.0-91-generic #92-Ubuntu SMP Fri Feb 28 11:09:48 UTC
```

```
2020 x86_64 x86_64 x86_64 GNU/Linux
root@blockchain:/usr/local/bin# lsb_release -a
No LSB modules are available.
Distributor ID: Ubuntu
Description:    Ubuntu 18.04.2 LTS
Release:        18.04
Codename:       bionic
root@blockchain:/usr/local/bin#
```

当前操作系统是Ubuntu 18.04.2 LTS版本，Docker采用19.03版本，满足安装Compose使用。

2. 安装 Docker-Compose 工具

```
root@blockchain:~# apt install docker-compose
Reading package lists... Done
Building dependency tree
Reading state information... Done
...（此处省略了部分显示信息）
Setting up docker-compose (1.17.1-2) ...
Processing triggers for man-db (2.8.3-2ubuntu0.1) ...
```

3. 测试和查看 Compose 版本

```
root@blockchain:~# docker-compose version
docker-compose version 1.17.1, build unknown
docker-py version: 2.5.1
CPython version: 2.7.17
OpenSSL version: OpenSSL 1.1.1  11 Sep 2018
```

从命令显示的信息看，docker-compose及其需要的依赖包已经安装成功，下面开始部署高可用Web服务。

四、部署高可用 Web 服务

1. 创建 compose 文件夹

```
root@blockchain:~# mkdir compose
```

2. 编辑 docker-compose.yml 文件

```
root@blockchain:~# cd compose/
root@blockchain:~/compose# vi docker-compose.yml
```

输入以下内容：

```
version: '2'
services:
    web1:
            image: nginx
            expose:
                - 80
            volumes:
                - ./web1:/usr/share/nginx/html
```

```
    web2:
        image: nginx
        expose:
            - 80
        volumes:
            - ./web2:/usr/share/nginx/html
    haproxy:
        image: haproxy
        volumes:
            - ./haproxy/haproxy.cfg:/usr/local/etc/haproxy/haproxy.cfg:ro
        links:
            - web1
            - web2
        ports:
            - "80:80"
        expose:
            - "80"
```

这里对几个常用命令进行解释：

version：'3'声明该docker-compose.yml版本号，不同的版本命令有一定的差异，可以参考官网https://docs.docker.com/compose/compose-file相关信息。

services：声明提供的服务，一般是由多个容器提供。

image：声明启动容器的镜像。

expose：声明暴露的端口，是容器内部端口，并不与宿主机的端口关联。

volumes：声明挂载的卷，格式是（HOST:CONTAINER）或加上访问模式（HOST: CONTAINER: ro）。

links：链接到其他服务中的容器，使用服务名称，或者服务名称：服务别名格式（SERVICE: ALIAS）。

ports：声明暴露的端口，使用宿主机:容器（HOST:CONTAINER）格式，或者仅仅指定容器的端口（宿主机将会随机选择端口）。

3. 创建 web1 和 web2 文件夹

```
root@blockchain:~/compose# mkdir web1
root@blockchain:~/compose# mkdir web2
```

4. 编写测试页面

```
root@blockchain:~/compose# ls
docker-compose.yml  web1  web2
root@blockchain:~/compose# echo "web1web1web1web1web1" >web1/index.html
root@blockchain:~/compose# echo "web2web2web2web2web2" >web2/index.html
```

5. 下载 nginx 和 haproxy 镜像

```
root@blockchain:~/compose# docker pull nginx
Using default tag: latest
```

```
latest: Pulling from library/nginx
68ced04f60ab: Already exists
28252775b295: Pull complete
a616aa3b0bf2: Pull complete
Digest: sha256:2539d4344dd18e1df02be842ffc435f8e1f699cfc55516e2cf2cb16b7a9aea0b
Status: Downloaded newer image for nginx:latest
docker.io/library/nginx:latest
root@blockchain:~/compose# docker pull haproxy
Using default tag: latest
latest: Pulling from library/haproxy
68ced04f60ab: Already exists
48983962b8f5: Pull complete
382fb8a7e67f: Pull complete
Digest: sha256:18d287a2191c4ae3309a5243f9d80aed4a286f00d0259485e24a3dd3658751aa
Status: Downloaded newer image for haproxy:latest
docker.io/library/haproxy:latest
```

6. 配置haproxy负载均衡

（1）创建haproxy文件夹

root@blockchain:~/compose# mkdir haproxy

（2）编辑haproxy.cfg文件

root@blockchain:~/compose# vi haproxy/haproxy.cfg

输入如下内容：

```
global
        log 127.0.0.1 local0
        log 127.0.0.1 local1 notice
defaults
        log global
        mode http
        option httplog
        option dontlognull
        timeout connect 5000ms
        timeout client 50000ms
        timeout server 50000ms
        stats uri /status
frontend balancer
        bind 0.0.0.0:80
        default_backend web_backends
backend web_backends
        balance roundrobin
        server web1 web1:80 check
        server web2 web2:80 check
```

7. 清除宿主机中运行的容器

为了避免在操作中出现端口占用等情况，需要先把正在运行的或者已经退出的容器清除。

注意：在测试学习环境中可以使用该命令，在生产环境中绝对不能使用该命令。

```
root@blockchain:/dockerfile# docker rm  -f $(docker ps -aq)
```

8. 执行 docker-compose 命令运行容器

```
root@blockchain:~/compose# docker-compose up
Starting compose_web2_1 ... done
Starting compose_web1_1 ... done
Starting compose_haproxy_1 ... done
Attaching to compose_web2_1, compose_web1_1, compose_haproxy_1
haproxy_1  | [NOTICE] 083/133328 (1) : New worker #1 (6) forked
```

9. 测试

在浏览器中输入宿主机的IP地址进行访问测试，图5-2所示为成功访问界面。

至此，测试成功，正常访问Nginx提供的Web服务。

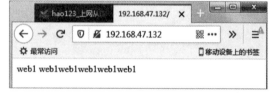

图 5-2 网页测试

任务总结

① 本任务的网络架构服务前端通过haproxy负载均衡器实现Web服务，后端用两个Nginx容器实例提供Web服务，在实际工作中，这种架构并不能提供真正的高可用Web服务，即使后端增加Nginx容器实例也不能实现高可用Web服务服务，主要原因是所有的服务均在同一个宿主机中运行，是典型的单点故障设计，要想真正实现高可用，需要多个宿主机的冗余设计，这需要在以后学习多主机环境下的swarm或者kubernetes编排实现。

② docker-compose.yml文件的编写有严格的缩进格式要求，而且采用空格进行缩进，不能使用【Tab】键进行缩进。

③ docker-compose.yml文件的命令随着版本的不同有所不同，可以参考官网信息。

④ 可以用docker-compose --help查看常用命令，尝试对容器进行操作管理。

任务扩展

① 查看运行的容器，尝试关闭其中某一个Nginx容器，然后测试Web服务，测试是否能正常持续提供Web服务（系统是提供高可用的，关闭任何一个Nginx容器是可以正常访问网页的）。

② 阅读如下附件，了解haproxy（参考:http://www.tianfeiyu.com）。

任务二　使用 Compose 实现个人博客网站建设

任务场景

由于业务需求，安安公司需要部署个人博客网站，计划采用WordPress实现，现需要技术人员采用docker-compose编排技术，实现个人博客网站建设。

任务描述

本任务学习采用docker-compose编排技术，实现个人博客网站建设。

任务目标

◎ 了解WordPress个人博客软件
◎ 能够根据WordPress个人博客软件部署需求编写docker-compose配置文件
◎ 能够使用命令编排和测试WordPress个人博客网站

任务实施

WordPress是使用PHP语言开发的博客平台，用户可以在支持PHP和MySQL数据库的服务器上架设属于自己的网站。也可以把WordPress当作一个内容管理系统（CMS）来使用。

WordPress是一款个人博客系统，并逐步演化成一款内容管理系统软件，它是使用PHP语言和MySQL数据库开发的。用户可以在支持PHP和MySQL数据库的服务器上使用自己的博客。

WordPress有许多第三方开发的免费模板，安装方式简单易用。要做一个自己的模板，需要有一定的专业知识。如至少要懂得标准通用标记语言下的一个应用HTML代码、CSS、PHP等相关知识。

WordPress官方支持中文版，有爱好者开发的第三方中文语言包，如wopus中文语言包。WordPress拥有成千上万个各式插件和不计其数的主题模板样式。

一、环境设计

1. 架构设计

采用简单的博客网站架构设计如图5-3所示，WordPress由两个容器组成，分别是WordPress和mysql。两个容器的命名也分别用WordPress和mysql实现。

2. 目录结构设计

在管理员目录下创建WordPress目录，创建如下目录结构。

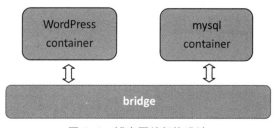

图5-3 博客网站架构设计

```
root@blockchain:~/wordpress# mkdir db
root@blockchain:~/wordpress# mkdir wp
root@blockchain:~/wordpress# cd ..
root@blockchain:~# tree wordpress/
wordpress/
├── db
└── wp
2 directories, 0 files
```

二、编辑配置文件

按照如下内容在wordpress目录下编辑docker-compose.yml文件。

```
root@blockchain:~/wordpress# vi docker-compose.yaml
#指定版本
version: '3'
#指定服务，本部分包含db和WordPress服务
services:
#    定义db容器
  db:
    image: mysql:5.7
    volumes:
        - ./db:/var/lib/mysql
    restart: always
    ports:
        - 3306
    environment:
        MYSQL_ROOT_PASSWORD: wordpress
        MYSQL_DATABASE: wordpress
        MYSQL_USER: wordpress
        MYSQL_PASSWORD: wordpress
    networks:
        - wp_net
#定义wordpress容器
  wordpress:
    depends_on:
            - db
    image: wordpress
    ports:
        - "80:80"
    restart: always
    environment:
        WORDPRESS_DB_HOST: db:3306
        WORDPRESS_DB_USER: wordpress
        WORDPRESS_DB_PASSWORD: wordpress
        WORDPRESS_DB_NAME: wordpress
    volumes:
        - ./wp:/var/www/html/wp/
    networks:
        - wp_net
#声明网络
networks:
    wp_net:
        driver: bridge
```

命令说明：

depends_on：用来声明依赖，也就是wordpress容器的启动需要定义的db服务支持。

networks：定义项目中的网络，名称为wp_net，driver为bridge，此处是先定义，然后在

service定义的服务中才能使用。

environment：定义了MySQL的环境变量，此处必须定义的有MYSQL_ROOT_PASSWORD: wordpress和MYSQL_DATABASE: wordpress环境变量。

restart：always定义如果容器启动失败会每隔10 s再次启动。

三、启动和测试

1. 启动网站

```
root@blockchain:~/wordpress# docker-compose up
Starting wordpress_db_1 ... done
Starting wordpress_wordpress_1 ... done
Attaching to wordpress_db_1, wordpress_wordpress_1
...
```

2. 测试

（1）查看容器运行状况

```
root@blockchain:~/wordpress# docker-compose ps
       Name                    Command             State          Ports
---------------------------------------------------------------------------
 wordpress_db_1         docker-entrypoint.sh mysqld    Up     0.0.0.0:32786-
>3306/tcp, 33060/tcp
 wordpress_wordpress_1  docker-entrypoint.sh apach ... Up     0.0.0.0:80-
>80/tcp
```

（2）浏览器测试

在浏览器中输入192.168.47.132宿主机的IP地址，打开欢迎界面，如图5-4所示，输入站点标题、用户名等信息，然后单击"安装WordPress"按钮。

图5-4　欢迎界面

进入设置完成界面,如图5-5所示。

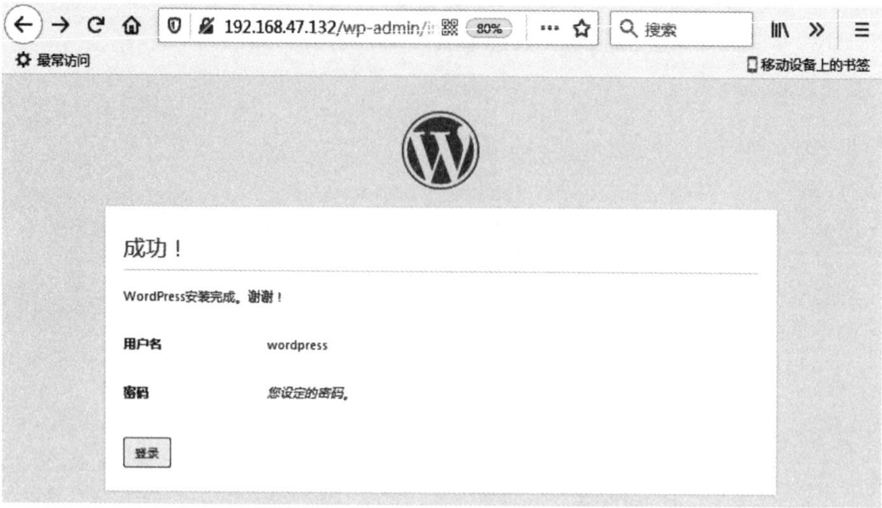

图 5-5　设置完成界面

单击"登录"按钮,进入登录界面,如图5-6所示。

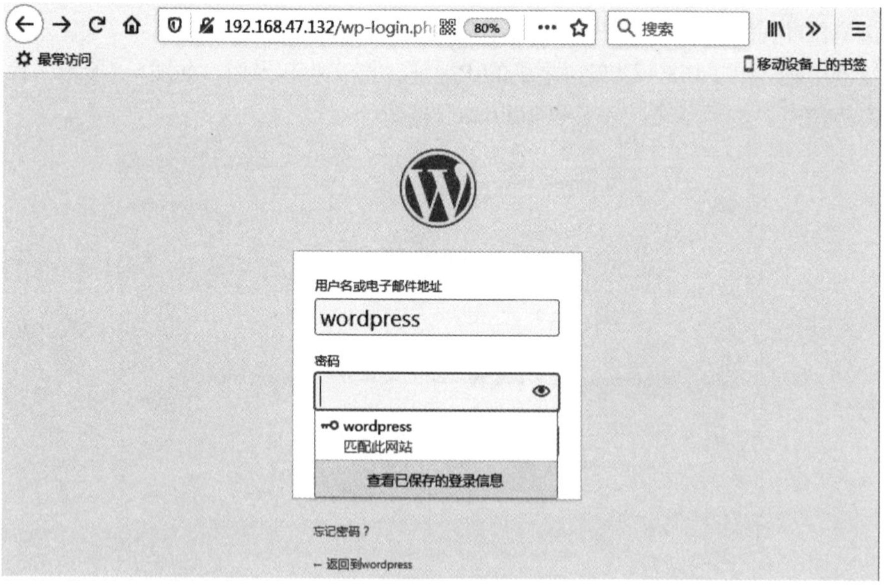

图 5-6　登录界面

输入用户名和密码进行登录,进入自己的博客设置环境,如图5-7所示。

项目五 使用 Compose 编排服务

图 5-7 博客设置环境界面

至此,博客网站安装完成。

任务总结

① 个人博客网站需要数据库的支持,架构设计采用WordPress和MySQL实现,成本较小。

② 编制docker-compose配置文件,需要注意WordPress和MySQL的关系,这里使用了depends_on命令,也就是WordPress需要依赖MySQL的启动。

③ docker-compose配置文件networks是在项目中先定义,然后在service定义的服务中才能使用,如果没有定义则不能在service服务中使用。如果是简单的项目,可以不进行networks定义,

115

在service服务中也不需要引用定义的网络，采用默认的bridge网络即可。

④ docker-compose配置文件restart: always定义如果容器启动失败会每隔10 s再次启动，还有其他一些选项，如restart: "no"，restart: on-failure，restart: unless-stopped等，用户可以根据需要进行选择。

任务扩展

在本任务中，读者可以尝试修改docker-compose配置文件，将网络networks定义删除，采用默认的bridge网络，查看容器的IP地址，并与本任务的配置进行对比。

任务三 运用 Compose 使用现有镜像配置 LNMP 网站

任务场景

由于业务需求，安安公司需要部署LNMP网站，为了顺利进行，现需要技术人员运用docker-compose编排技术进行测试部署Lnmp网站，测试成熟后进行生产发布。

任务描述

本任务学习采用docker-compose编排技术部署LNMP网站。

任务目标

◎ 了解LNMP网站架构。
◎ 能够根据LNMP网站架构部署需求编写docker-compose配置文件。
◎ 能够使用命令编排和测试LNMP网站架构。

任务实施

一、环境设计

1. 架构设计

采用在宿主机上利用bridge网络部署，LNMP架构设计如图5-8所示。LNMP由3个容器组成，分别是nginx、php和mysql。

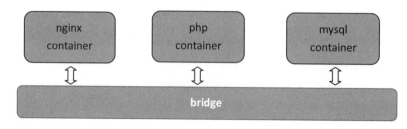

图5-8 LNMP 架构设计

2. 目录结构设计

在管理员目录下创建lnmp目录，并逐步创建如下目录结构和编辑相关文档。

```
root@blockchain:/home/adminroot/lnmp# tree
.
├── docker-compose.yml
├── mysql
│   ├── conf
│   │   └── my.cnf
│   └── data
├── nginx
│   └── nginx.conf
└── wwwroot
    └── index.php
```

二、编辑配置文件

需要编辑docker-compose.yml文件、nginx的配置文件nginx.conf、mysql的配置文件my.cnf和测试网页文件。

1. 编辑 docker-compose.yml 文件

按照如下内容编辑docker-compose.yml文件，带#的行是对配置的解释。

```yaml
# 指定服务版本号
version: '3'
# 服务
services:
# 定义nginx服务名称
  nginx:
  # 指定服务容器名字
    hostname: nginx
  # 使用镜像
    image: nginx
  # 映射数组级的端口
    ports:
            - 80:80
            - 443:443
    restart: always
    depends_on:
            - mysql
            - php
  # 映射php文件服务别名
    links:
            - php:php-cgi
  # 映射服务数据卷路径
    volumes:
            - ./wwwroot:/usr/share/nginx/html
```

```yaml
            - ./nginx/nginx.conf:/etc/nginx/nginx.conf
      networks:
            - lnmp
  # 服务名称
    php:
    # 指定服务容器名字
        hostname: php
    # 使用的镜像
        image: bitnami/php-fpm
        # 映射mysql服务别名
        ports:
  ports:
            - 9000:9000
        expose:
            - "9000"
        links:
            - mysql:mysql-db
    # 映射服务数据卷路径
        volumes:
            - ./wwwroot:/usr/share/nginx/html
        networks:
            - lnmp
        stdin_open: true
        tty: true
  # 服务名称
    mysql:
      # 指定服务容器名字
        hostname: mysql
      # 指定使用官方mysql5.6版本
        image: mysql
      # 映射端口
        ports:
            - 3306:3306
      # 映射服务数据卷路径
        volumes:
            - ./mysql/conf:/etc/mysql/conf.d
            - ./mysql/data:/var/lib/mysql
      # 指定数据库变量
        environment:
        # 设置数据库密码，这里用123456，用户根据需要更改
            MYSQL_ROOT_PASSWORD: 123456
        # 添加user用户
            MYSQL_USER: user
        # 设置user用户密码
            MYSQL_PASSWORD: user123
```

```
        networks:
            - lnmp
#声明网络
networks:
    lnmp:
        driver: bridge
```

2. 编辑 nginx.conf 文件

在nginx文件夹中编辑nginx.conf文件，重点是修改location中的内容，更改nginx和PHP的连接配置，在默认配置文件中注释掉了，这里按照如下进行更改（蓝色显示）。

```
root@blockchain:/home/adminroot/lnmp/nginx# vi nginx.conf
...    （此处省略了部分显示信息）
    server {
        listen 80;
        server_name localhost;
        #指定网页所在的目录
        root /usr/share/nginx/html;
        index index.html index.php;
        location ~ .*\.php$ {
            #指明php套接字，其中php-cgi是docker-compose.yml文件中定义的PHP别名。
            fastcgi_pass php-cgi:9000;
            fastcgi_index index.php;       #指明网页文件
            #以下两行指明PHP脚本文件名所在的路径和文件名，其中include是参数定义。
            fastcgi_param SCRIPT_FILENAME /usr/share/nginx/html/$fastcgi_script_name;
            include fastcgi_params;
        }
    }
    #gzip  on;
    include /etc/nginx/conf.d/*.conf;
}
```

3. 编辑网页 index.php 文件

在wwwroot文件夹中编辑index.php网页文件，本次测试打印PHP基本信息。

```
root@blockchain:/home/adminroot/lnmp/wwwroot# vi index.php
<?php phpinfo(); ?>
```

4. 编辑 mysql 配置文件

该配置文件更改后映射到mysql容器中，目的是当需要修改时，直接在宿主机上修改就可以了，方便配置变更，也可以直接用mysql默认配置文件，不影响测试网站使用。

```
root@blockchain:/home/adminroot/lnmp/mysql/conf# vi my.cnf
[mysqld]
user=mysql
port=3306
datadir=/var/lib/mysql
```

```
socket=/var/run/mysqld/mysqld.sock
pid-file=/var/run/mysqld/mysqld.pid
#log_error=/var/log/mysql/error.log
character_set_server = utf8
max_connections=3600
```

三、利用 docker-compose 部署 LNMP 和测试

1. 执行 docker-compose 命令一键部署 LNMP

该命令第一次执行时显示信息较多，下面用省略号省略了部分显示信息。

```
root@blockchain:/home/adminroot/lnmp# docker-compose up
...
Creating lnmp_mysql_1 ... done
Creating lnmp_php_1   ... done
Creating lnmp_nginx_1 ... done
Attaching to lnmp_mysql_1, lnmp_php_1, lnmp_nginx_1
...
mysql_1  | 2020-04-21T10:24:15.515104Z 0 [System] [MY-011323] [Server]
X Plugin ready for connections. Socket: '/var/run/mysqld/mysqlx.sock'
bind-address: '::' port: 33060
```

2. 查看运行状况

```
root@blockchain:/home/adminroot/lnmp# docker-compose ps
     Name              Command                  State              Ports
-----------------------------------------------------------------------------
 lnmp_mysql_1   docker-entrypoint.sh mysqld     Up      0.0.0.0:3306->3306/
tcp, 33060/tcp
 lnmp_nginx_1   nginx -g daemon off;            Up      0.0.0.0:443->443/
tcp, 0.0.0.0:80->80/tcp
 lnmp_php_1     php-fpm -F --pid /opt/bitn ...  Up      0.0.0.0:9000->9000/tcp
```

可以看到3个容器正常运行，都处于Up状态，端口映射信息正常，说明已经成功运行。

3. 进行网页测试

在浏览器中输入宿主机的IP地址测试，图5-9所示调用了phpinfo()显示PHP信息，就表示成功。

4. 关闭容器

使用docker-compose down命令关闭容器。

```
root@blockchain:/home/adminroot/lnmp# docker-compose down
Stopping lnmp_nginx_1 ... done
Stopping lnmp_php_1   ... done
Stopping lnmp_mysql_1 ... done
Removing lnmp_nginx_1 ... done
Removing lnmp_php_1   ... done
Removing lnmp_mysql_1 ... done
Removing network lnmp_lnmp
```

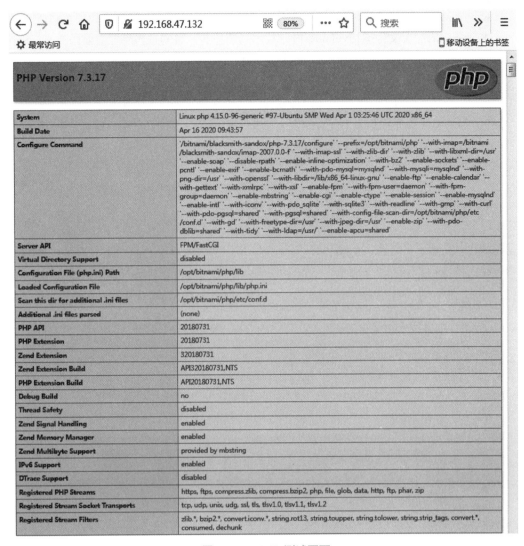

图 5-9　LNMP 测试页面

任务总结

至此，利用docker-compose实现了部署LNMP容器，该方案使用现有官方的nginx、php和mysql镜像进行一键部署，关键内容如下：

① 编写docker-compose.yaml文件时，语法需要严格按照yaml文件格式进行编写，注意相同的缩进关系。本任务中定义3个服务，3个服务的缩进必须相同，服务内部的键值对在本服务中对齐即可。

② 确定服务之间的关系，编写docker-compose.yaml文件时，进行编排nginx、php和mysql这几个服务，并根据内在的关系通过links和depends_on命令进行连接和依赖关系处理，确定它们依赖和启动先后顺序。

③ 对于nginx的配置文件重点是PHP的参数、套接字配置、cgi参数等，配置错误会导致nginx不能解析php，导致访问网页失败。在这里强调的是php套接字定义，其中php-cgi是docker-compose.yml文件中用links定义的PHP别名，如果引用错误，会出现找不到php-cgi主机的错误。

④ 对于mysql配置，注意环境变量MYSQL_ROOT_PASSWORD参数设置，该项是必须设置的。

⑤ 本任务中的配置文件运用volumes进行数据卷的挂载，目的是当配置需要修改时，修改配置参数比较方便，在宿主机中进行修改，重新更新配置就可以生效。

任务扩展

本任务用的容器镜像都是采用现有的官方镜像，用户可以尝试自己用原始官方ubuntu镜像通过安装软件的方式生成新的镜像完成LNMP网站的部署。

任务四　使用 Compose 编译实现 LNMP 网站建设

任务场景

由于业务需求，安安公司需要部署LNMP网站，计划采用编译的方式定制自己的镜像，运用docker-compose编排技术测试实现。

任务描述

本任务学习采用docker-compose编排技术，采用编译的方式部署LNMP网站。

任务目标

◎ 熟悉Linux环境中编译的基本命令
◎ 能够根据LNMP网站部署需求编写dockerfile配置文件
◎ 能够根据LNMP网站架构部署需求编写docker-compose配置文件
◎ 能够使用命令编排和测试LNMP网站架构

任务实施

一、环境设计

1. 架构设计

采用图5-10所示架构，LNMP由3个容器组成，分别是nginx、php和mysql。3个容器的命名也分别用nginx、php和mysql实现，基础镜像采用ubuntu镜像。

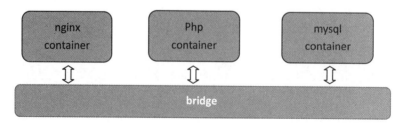

图 5-10　LNMP 网站架构

2. 目录结构设计

在管理员目录下创建compose_lnmp目录，并逐步创建如下目录结构：

```
[root@localhost ~]# tree compose_lnmp/
compose_lnmp/
├── docker-compose.yml
├── mysql
│   ├── conf
│   │   └── my.cnf
│   └── data
├── nginx
│   ├── Dockerfile
│   ├── nginx-1.12.1.tar.gz
│   └── conf
│       └── nginx.conf
├── php
│   ├── Dockerfile
│   ├── php-5.6.31.tar.gz
│   └── php.ini
└── wwwroot
    └── test.php
```

二、安装配置 mysql、nginx 和 php 服务

1. 下载和配置基础镜像

下载ubuntu镜像，将为nginx和php服务提供基础镜像，并对其源改换成阿里云的源，安装vim、inetutils-ping和net-tools工具，以便于日后维护。

（1）下载ubuntu镜像

```
root@blockchain:~# docker pull ubuntu
Using default tag: latest
latest: Pulling from library/ubuntu
5bed26d33875: Pull complete
f11b29a9c730: Pull complete
930bda195c84: Pull complete
78bf9a5ad49e: Pull complete
Digest: sha256:bec5a2727be7fff3d308193cfde3491f8fba1a2ba392b7546b43a051853a341d
Status: Downloaded newer image for ubuntu:latest
docker.io/library/ubuntu:latest
```

（2）运行ubuntu容器

```
root@blockchain:~# docker run -it ubuntu
root@04e996303ba8:/#
```

（3）更改阿里云的源

```
root@04e996303ba8:/# echo 'deb http://mirrors.aliyun.com/ubuntu/bionic
```

```
main restricted universe multiverse
    > deb-src http://mirrors.aliyun.com/ubuntu/bionic main restricted
universe multiverse
    > deb http://mirrors.aliyun.com/ubuntu/bionic-security main restricted
universe multiverse
    > deb-src http://mirrors.aliyun.com/ubuntu/bionic-security main
restricted universe multiverse
    > deb http://mirrors.aliyun.com/ubuntu/bionic-updates main restricted
universe multiverse
    > deb-src http://mirrors.aliyun.com/ubuntu/bionic-updates main restricted
universe multiverse
    > deb http://mirrors.aliyun.com/ubuntu/bionic-backports main restricted
universe multiverse
    > deb-src http://mirrors.aliyun.com/ubuntu/bionic-backports main
restricted universe multiverse
    > deb http://mirrors.aliyun.com/ubuntu/bionic-proposed main restricted
universe multiverse
    > deb-src http://mirrors.aliyun.com/ubuntu/bionic-proposed main
restricted universe multiverse' > /etc/apt/sources.list
```

（4）进行更新

```
root@04e996303ba8:/# apt update
Get:1 http://mirrors.aliyun.com/ubuntu bionic InRelease [242 kB]
… （此处省略了部分信息）
Fetched 32.5 MB in 4s (7466 kB/s)
Reading package lists... Done
Building dependency tree
Reading state information... Done
14 packages can be upgraded. Run 'apt list --upgradable' to see them.
```

（5）安装vim、inetutils-ping和net-tools工具

```
root@04e996303ba8:/# apt install -y vim inetutils-ping net-tools
…（此处省略了log信息）
```

（6）测试ping和ifconfig命令工具功能

```
root@04e996303ba8:/# ping -c 2 www.baidu.com
PING www.a.shifen.com (14.215.177.39): 56 data bytes
64 bytes from 14.215.177.39: icmp_seq=0 ttl=127 time=8.850 ms
64 bytes from 14.215.177.39: icmp_seq=1 ttl=127 time=8.579 ms
--- www.a.shifen.com ping statistics ---
2 packets transmitted, 2 packets received, 0% packet loss
round-trip min/avg/max/stddev = 8.579/8.715/8.850/0.136 ms
root@04e996303ba8:/# ifconfig
eth0: flags=4163<UP,BROADCAST,RUNNING,MULTICAST>  mtu 1500
        inet 172.17.0.2  netmask 255.255.0.0  broadcast 172.17.255.255
```

```
            ether 02:42:ac:11:00:02   txqueuelen 0   (Ethernet)
            RX packets 32533  bytes 47435986 (47.4 MB)
            RX errors 0  dropped 0  overruns 0  frame 0
            TX packets 16037  bytes 879309 (879.3 KB)
            TX errors 0  dropped 0 overruns 0  carrier 0  collisions 0
    lo: flags=73<UP,LOOPBACK,RUNNING>  mtu 65536
            inet 127.0.0.1  netmask 255.0.0.0
            loop  txqueuelen 1000  (Local Loopback)
            RX packets 0  bytes 0 (0.0 B)
            RX errors 0  dropped 0  overruns 0  frame 0
            TX packets 0  bytes 0 (0.0 B)
            TX errors 0  dropped 0 overruns 0  carrier 0  collisions 0
```

经过运行工具测试，ping和ifconfig命令工具都能正常工作。

（7）将该容器打包成新的镜像

用docker commit命令将该容器打包成镜像，命名为ubuntu_with_vimnet。

先查看容器的ID，然后打包。

```
root@blockchain:~# docker ps
CONTAINER ID    IMAGE     COMMAND       CREATED         STATUS        PORTS       NAMES
04e996303ba8    ubuntu    "/bin/bash"   19 minutes ago  Up 19 minutes             zen_wright
root@blockchain:~#docker commit 04e996303ba8 ubuntu_with_vimnet
sha256:366b039c277a7aaa905c69c8f1a323fb95980a09b84100b3533ca408b55da3a6
```

（8）下载mysql镜像

下载mysql镜像，提供mysql容器服务。

```
root@blockchain:~/compose-lnmp# docker pull mysql
…（此处省略log输出）。
```

2. 配置 Nginx 服务

（1）创建lnmp项目工作目录

创建compose-lnmp目录作为LNMP的项目管理目录，然后创建nginx容器的配置目录。

```
root@blockchain:~# mkdir compose-lnmp
root@blockchain:~# cd compose-lnmp
root@blockchain:~/compose-lnmp# mkdir nginx
root@blockchain:~/compose-lnmp# cd nginx
root@blockchain:~/compose-lnmp/nginx#
```

另外创建一个存放网页的文件夹，随后将测试网页存放在这里。

```
root@blockchain:~/compose-lnmp# mkdir wwwroot
```

（2）下载nginx源包

到https://nginx.org/en/download.html网站找到下载nginx链接，复制链接，然后用wget命令进行下载。本项目使用https://nginx.org/download/nginx-1.17.9.tar.gz源包。

```
root@blockchain:~/compose-lnmp/nginx# wget https://nginx.org/download/
```

```
nginx-1.17.9.tar.gz
    --2020-03-26 13:04:39--  https://nginx.org/download/nginx-1.17.9.tar.gz
    Resolving nginx.org (nginx.org)... 62.210.92.35, 95.211.80.227,
2001:1af8:4060:a004:21::e3
    Connecting to nginx.org (nginx.org)|62.210.92.35|:443... connected.
    HTTP request sent, awaiting response... 200 OK
    Length: 1039136 (1015K) [application/octet-stream]
    Saving to: 'nginx-1.17.9.tar.gz.1'
    nginx-1.17.9.tar.gz.1   100%[=======================>]   1015K  13.1KB/s
in 86s
    2020-03-26 13:06:13(11.7 KB/s)-'nginx-1.17.9.tar.gz.1'saved [1039136/1039136]
```

查看下载后的nginx数据包。

```
root@blockchain:~/compose-lnmp/nginx# ls
nginx-1.17.9.tar.gz
```

（3）创建nginx镜像dockerfile文件

创建nginx镜像dockerfile文件，输入如下内容，带#的行为解释行，已经解释很清楚了，这里不再对dockerfile文件进一步解释。

```
root@blockchain:~/compose-lnmp/nginx# vi dockerfile
# 指定镜像
FROM ubuntu-with-vimnet
# 指定管理人员，或者说明
MAINTAINER ubuntu with net-tools and inetutils-ping by yangjianqing
# 执行命令安装编译库文件
RUN apt install -y gcc make openssl build-essential libtool libpcre3
libpcre3-dev zlib1g-dev
# 添加解压nginx包到/tmp目录下
ADD nginx-1.17.9.tar.gz /tmp
# 进入目录进行编译安装
RUN cd /tmp/nginx-1.17.9 && ./configure --prefix=/usr/local/nginx && make -
j 2 && make install &&echo "PATH=/usr/local/nginx/bin:/usr/local/nginx/sbin:/
usr/local/nginx/conf:$PATH" >> /etc/profile
# 删除容器内置配置文件
RUN rm -f /usr/local/nginx/conf/nginx.conf
# 复制本地配置文件到容器内
COPY ./conf/nginx.conf /usr/local/nginx/conf
# 声明暴露端口
EXPOSE 80
# 启动容器Nginx服务，指定全局命令daemon off保证服务在前台运行不会关闭
CMD ["/usr/local/nginx/sbin/nginx", "-g", "daemon off;"]
```

（4）配置nginx的配置文件

nginx.conf作为nginx的配置文件，在编译安装nginx的过程中会自动产生，但是为了方便管理，在宿主机创建该文件，通过卷挂载的方式挂载到nginx容器中，方便在需要时进行更改维护，

编辑nginx.conf文件，输入如下内容：

```
root@blockchain:~/compose-lnmp/nginx# vi nginx.conf
user    root;
worker_processes   auto;
error_log   logs/error.log   info;
pid         logs/nginx.pid;

events {
    use epoll;
}
http {
    include       mime.types;
    default_type  application/octet-stream;
    log_format main '$remote_addr-$remote_user [$time_local] "$request" '
                    '$status $body_bytes_sent "$http_referer" '
                    '"$http_user_agent" "$http_x_forwarded_for"';
    access_log logs/access.log main;
    sendfile        on;
    keepalive_timeout   65;
    server {
        listen 80;
        server_name localhost;
        root html;
        index index.html index.php;
        location ~ \.php$ {
            root /usr/local/nginx/html
            fastcgi_pass php-cgi:9000;
fastcgi_index index.php;
            fastcgi_param SCRIPT_FILENAME/usr/local/nginx/html/$fastcgi_script_name;
            include fastcgi_params;
        }
    }
}
```

3. 配置 mysql 服务

mysql服务采用源镜像，不再自己打包镜像。

(1) 创建mysql相关目录

创建mysql目录，并在mysql目录下创建conf目录，作为主配置文件所在目录。

```
root@blockchain:~/compose-lnmp# mkdir mysql
root@blockchain:~/compose-lnmp# cd mysql/
root@blockchain:~/compose-lnmp/mysql# mkdir data
root@blockchain:~/compose-lnmp/mysql# mkdir conf
root@blockchain:~/compose-lnmp/mysql# cd conf
```

root@blockchain:~/compose-lnmp/mysql/conf#

(2) 配置mysql配置文件

在./mysql/conf文件夹下创建文件my.cnf，输入以下内容：

```
root@blockchain:~/compose-lnmp/mysql/conf# vi my.cnf
[mysqld]
user=mysql
port=3306
datadir=/var/lib/mysql
socket=/var/run/mysqld/mysqld.sock
pid-file=/var/run/mysqld/mysqld.pid
log_error=/var/log/mysql/error.log
character_set_server = utf8
max_connections=3600
```

4. 配置php服务

(1) 创建php相关目录

创建php目录。

root@blockchain:~/compose-lnmp# mkdir php

(2) 下载php源包

到https://www.php.net/downloads网站找到下载php链接，复制链接，然后用wget命令进行下载。本项目用https://www.php.net/distributions/php-7.4.4.tar.gz源包。

```
root@blockchain:~/compose-lnmp/php# wget https://www.php.net/distributions/php-7.4.4.tar.gz
--2020-03-27 02:45:45--https://www.php.net/distributions/php-7.4.4.tar.gz
Resolving www.php.net (www.php.net)... 185.85.0.29, 2a02:cb40:200::1ad
Connecting to www.php.net (www.php.net)|185.85.0.29|:443... connected.
HTTP request sent, awaiting response... 200 OK
Length: 16477200 (16M) [application/octet-stream]
Saving to: 'php-7.4.4.tar.gz'

php-7.4.4.tar.gz    100%[============================>]  15.71M  273KB/s   in 56s

2020-03-27 02:46:47(288 KB/s)-'php-7.4.4.tar.gz' saved[16477200/16477200]
```

(3) 创建Dockerfile文件

在php目录下创建Dockerfile文件，输入如下内容：

```
root@blockchain:~/compose-lnmp/php# vi dockerfile
# 指定镜像
FROM ubuntu_with_vimnet
# 指定系统管理员
MAINTAINER ubuntu with net-tools and inetutils-ping by yangjianqing
# 安装编译库
RUN apt update&&apt-get install -y gcc libzip-dev bison autoconf build-essential pkg-config git-core\
```

```
    libltdl-dev libbz2-dev libxml2-dev libxslt1-dev libssl-dev libicu-dev
libpspell-dev\
    libenchant-dev libmcrypt-dev libpng-dev libjpeg8-dev libfreetype6-dev
libmysqlclient-dev\
    libreadline-dev libcurl4-openssl-dev librecode-dev libsqlite3-dev
libonig-dev &&\
  rm -f /bin/sh && ln -s /bin/bash /bin/sh
  # 下载压缩包解压到指定目录
  ADD php-7.4.4.tar.gz /tmp/
  # 编译安装php，执行相关命令
  RUN cd /tmp/php-7.4.4 && \
  ./configure --prefix=/usr/local/php7 \
  --with-config-file-scan-dir=/usr/local/php7/etc/php.d \
  --with-config-file-path=/usr/local/php7/etc --enable-mbstring\
   --enable-zip --enable-bcmath --enable-pcntl --enable-ftp\
   --enable-xml --enable-shmop --enable-soap --enable-intl --with-openssl\
   --enable-exif --enable-calendar --enable-sysvmsg --enable-sysvsem\
   --enable-sysvshm --enable-opcache --enable-fpm --enable-session\
   --enable-sockets --enable-mbregex --enable-wddx --with-curl\
   --with-iconv --with-gd --with-jpeg-dir=/usr --with-png-dir=/usr\
   --with-zlib-dir=/usr --with-freetype-dir=/usr --enable-gd-jis-conv\
   --with-openssl --with-pdo-mysql=mysqlnd --with-gettext=/usr\
   --with-zlib=/usr --with-bz2=/usr --with-recode=/usr \
  --with-xmlrpc --with-mysqli=mysqlnd && \
  make -j 4 && make install && \
  cp/usr/local/php7/etc/php-fpm.conf.default/usr/local/php7/etc/php-fpm.
conf && \
  cp /usr/local/php7/etc/php-fpm.d/www.conf.default /usr/local/php7/etc/
php-fpm.d/www.conf && \
  cp php.ini-production /usr/local/php7/etc/php.ini && \
  cp ./sapi/fpm/init.d.php-fpm /etc/init.d/php-fpm && \
  chmod +x /etc/init.d/php-fpm && groupadd nobody &&\
  sed -i "s/127.0.0.1/0.0.0.0/g"/usr/local/php7/etc/php-fpm.d/www.conf &&\
  source /etc/profile
  # 声明暴露端口
  EXPOSE 9000
  #启动php
  CMD /etc/init.d/php-fpm start && tail -F /var/log
```

三、编排和测试

1. 创建和编辑 docker-compose 配置文件

编辑docker-compose文件时有严格的缩进关系，在编辑的过程中必须遵守。

```
root@blockchain:~/compose-lnmp# vi docker-compose.yml
# 指定服务版本号
```

```yaml
version: '3'
# 服务
services:
# 服务名称
  nginx:
# 指定服务容器名字
    hostname: nginx
# 构建
    build:
# 指定目录上下文构建镜像
      context: ./nginx
# 指定dockerfile文件名称
      dockerfile: Dockerfile
# 映射数组级的端口
    ports:
      - "80:80"
# 映射php文件服务别名
    links:
      - "php:php-cgi"
# 映射服务数据卷路径
    volumes:
      - ./wwwroot:/usr/local/nginx/html
# 服务名称
  php:
# 指定服务容器名字
    hostname: php
# 构建
    build: ./php
# 映射mysql服务别名
    links:
      - mysql:mysql-db
# 映射服务数据卷路径
    volumes:
      - ./wwwroot:/usr/local/nginx/html
# 服务名称
  mysql:
# 指定服务容器名字
    hostname: mysql
# 指定使用官方mysql5.6版本
    image: mysql
# 映射端口
    ports:
      - "3306:3306"
# 映射服务数据卷路径
    volumes:
```

```
        - ./mysql/conf:/etc/mysql/conf.d
        - ./mysql/data:/var/lib/mysql
# 指定数据库变量
    environment:
# 设置数据库密码，这里用123456，用户根据需要更改
      MYSQL_ROOT_PASSWORD: 123456
# 添加user用户
      MYSQL_USER: user
# 设置user用户密码
      MYSQL_PASSWORD: user123
```

2. 执行 docker-compose 命令一键部署 LNMP

这里执行docker-compose up命令需要在docker-compose.yml文件所在的目录下执行。经过编译、镜像制作和运行容器，最后出现如下界面，时间的长短取决于个人宿主机的速度。

```
root@blockchain:~/compose-lnmp# docker-compose up
... （此处省略了部分显示信息）
Creating network "compose-lnmp_default" with the default driver
Creating compose-lnmp_mysql_1 ... done
Creating compose-lnmp_php_1   ... done
Creating compose-lnmp_nginx_1 ... done
Attaching to compose-lnmp_mysql_1, compose-lnmp_php_1, compose-lnmp_nginx_1
... （此处省略了部分显示信息）
```

使用 docker-compose ps命令查看运行结果，可以看到名字为容器名称、运行的命令、状态和端口几列信息，可以清楚看到几个容器的状态处于Up状态。

```
root@blockchain:~/compose-lnmp# docker-compose ps
Name                    Command                          State   Ports
---------------------------------------------------------------------------------
compose-lnmp_mysql_1    docker-entrypoint.sh mysqld      Up      0.0.0.0:3306->3306/tcp
compose-lnmp_nginx_1    /usr/local/nginx/sbin/ngin...    Up      0.0.0.0:80->80/tcp
compose-lnmp_php_1      /bin/sh -c /etc/init.d/php...    Up      9000/tcp
```

3. 创建测试网页

```
root@blockchain:~/compose-lnmp/wwwroot# echo "<?php phpinfo()?>" > index.php
```

4. 浏览器测试

在浏览器中输入宿主机的IP地址，测试网页，图5-11所示为正常显示。

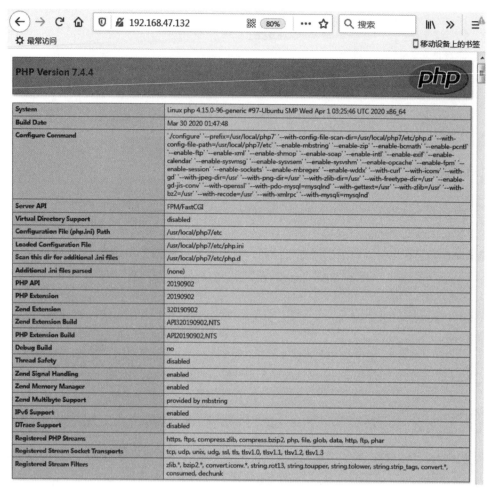

图 5-11 测试网页

至此，通过编译方式实现了LNMP网站的建设。

任务总结

本项目中通过编译的方式实现了lnmp网站的建设，重点关注编译、dockerfile文件和docker-compose文件的编写。

1. nginx 的 dockerfile 编写

① 编译部分注意，在dockerfile文件中通过ADD命令将下载的文件解压复制到指定地点，并用RUN命令启用编译命令进行编译。

② 替换配置文件，将nginx本身的配置文件删除，用自己编辑的配置文件替换。

2. nginx 配置文件编写

nginx的配置文件编写重点是location部分的php配置，需要正确配置。

3. php 的 dockerfile 编写

① php安装的组件比较多，参数多且易出错，配置时需要小心。

② --prefix、--with-config-file-scan-dir和--with-config-file-path=/usr/local/php7/etc参数是自定义的参数，编译后对应的文件位置是在此处定义的，如果随后需要进行调整，需要了解该

参数的配置。

③ 配置文件的复制，注意php-fpm.conf、www.conf&&和php.ini文件都是根据一些默认配置文件复制过来的。

④ www.conf文件监听地址的修改，需要将www.conf文件中的listen = 127.0.0.1:9000修改为listen = 0.0.0.0:9000。

⑤ 需要为php-fpm文件添加执行权限。

4. docker-compose 文件的编写

① docker-compose文件的编写需要注意几个数据卷的挂载。涉及nginx和php的网页主目录挂载，挂载到宿主目录中方便更改；mysql的数据库和配置文件挂载，方便数据的持久化和其他利用。

② 镜像build需要注意上下文的定义，在nginx中是当前目录的nginx文件夹，php中是当前目录的php文件夹，如果指定错误将不能正常生成镜像。

③ 数据库注意环境变量MYSQL_ROOT_PASSWORD变量设置。

任务扩展

本任务中的数据库镜像采用官方原始镜像实现，读者尝试采用本任务中的ubuntu原始镜像，通过下载数据库软件包，编译生成镜像方式提供mysql服务，然后测试lnmp网站。

项目六 认识和理解多主机网络

【项目综述】

公司的业务在一台主机上运行不具备高可用性,一旦节点故障,整个服务就瘫痪了。根据公司业务需求,安安公司需要提高公司业务服务的高可用性,需要研究多主机的运行。本项目主要学习多主机模式的网络架构。

【项目目标】

◎ 认识 docker-machine 架构
◎ 能够配置 docker-machine
◎ 能够描述 docker 网络的类型
◎ 能够叙述 overlay 网络结构
◎ 能够分析 overlay 网络桥接关系

任务一 安装配置 docker-machine

任务场景

为提升公司业务网络服务的可用性,安安公司决定采用多主机模式承载公司的网络业务,技术人员尝试采用docker-machine方式实现多主机之间的通信。

任务描述

本任务学习docker-machine,采用docker-machine方式实现多主机的通信,并部署docker容器测试网络通信。

任务目标

◎ 能够设计docker-machine配置环境
◎ 能够安装和配置docker-machine主机
◎ 能够将其他主机加入到docker-machine集群中
◎ 能够查看和测试docker-machine集群

任务实施

一、了解 docker-machine

docker-machine是一个简化安装Docker环境的工具。可以在虚拟主机上安装 Docker 的工具,并可以使用 docker-machine 命令来管理主机。

Docker Machine 也可以集中管理所有的 docker 主机，比如快速为 100 台服务器安装 docker。

Docker Machine 管理的虚拟主机可以是本地主机上的，也可以是云供应商，如阿里云、腾讯云、AWS或DigitalOcean。

使用 docker-machine 命令，可以启动，检查，停止和重新启动托管主机，也可以升级Docker 客户端和守护程序，以及配置 Docker 客户端与主机进行通信。

二、设计 docker-machine 部署环境

docker-machine工作在多主机环境，其中至少一台作为主控节点，称为master主机，其他主机作为master管理下的docker-machine客户节点。下面采用3台主机，分别是blockchain主机（主机IP为192.168.47.132）作为master管理节点，ubuntu-chain0（主机IP为192.168.47.137）和ubuntu-chain1（主机IP为192.168.47.139）作为docker-machine的客户节点。docker-machine网络架构如图6-1所示。

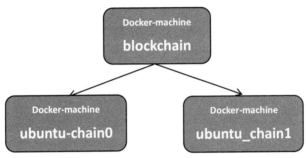

图 6-1　docker-machine 网络架构

三、下载 docker-machine 软件

```
root@blockchain:~# curl -L https://github.com/docker/machine/releases/download/v0.16.0/docker-machine-$(uname -s)-$(uname -m) >/tmp/docker-machine &&
> sudo mv /tmp/docker-machine /usr/local/bin/docker-machine &&
> chmod +x /usr/local/bin/docker-machine
  % Total    % Received % Xferd  Average Speed   Time    Time     Time  Current
                                 Dload  Upload   Total   Spent    Left  Speed
100   617  100   617    0     0     96      0  0:00:06  0:00:06 --:--:--   159
  1 26.8M    1  373k    0     0    318      0 24:36:07  0:20:01 24:16:06  3583
```

四、测试 docker-machine

1. 查看 docker-machine 版本

```
root@blockchain:~# docker-machine version
docker-machine version 0.16.0, build 702c267f
```

2. 查看当前 docker-machine 管理的主机

```
root@blockchain:~# docker-machine ls
NAME   ACTIVE   DRIVER   STATE   URL   SWARM   DOCKER   ERRORS
```

从查看结果上看，当前没有docker-machine客户主机。

五、下载补全命令脚本

下载补全命令脚本的目的是在完成docker-machine客户主机加入之后,在管理过程中提示符会显示管理某台docker-machine客户主机,这样管理起来比较方便。下载补全命令脚本并不是必需步骤。

1. 创建目录

```
root@blockchain:/usr/local/etc# mkdir bash_completion.d
```

2. 编辑下载脚本

```
root@blockchain:/usr/local/etc# vi .bash
```

输入:

```
base=https://raw.githubusercontent.com/docker/machine/v0.16.0
for i in docker-machine-prompt.bash docker-machine-wrapper.bash docker-machine.bash
do
    sudo wget "$base/contrib/completion/bash/${i}" -P /usr/local/etc/bash_completion.d
Done
```

3. 执行下载脚本

```
root@blockchain:/usr/local/etc# ./.bash
```

下载完成后可以看到下载的文件。

```
root@blockchain:/usr/local/etc/bash_completion.d# ls
docker-machine.bash   docker-machine-prompt.bash   docker-machine-wrapper.bash
```

加载运行docker-machine-prompt.bash。

```
root@blockchain:/usr/local/etc/bash_completion.d# source ./docker-machine-prompt.bash
```

4. 更改变量文本

将PS1='[\u@\h \W$(__docker_machine_ps1)]\$ '变量添加到$HOME/.bashrc文件中。

六、添加 machine 客户主机

docker-machine通过ssh连接到Docker主机,从网上下载并安装docker工具,需要用root权限来安装。在ubuntu系统下,默认禁止root用户登录系统,因此需要先配置root允许SSH登录系统并免交互登录或指定私钥登录。

本任务将ubuntu-chain1(主机IP为192.168.47.139)主机添加到docker-machine中。

1. 允许 root ssh 登录

在docker-machine客户主机ubuntu-chain1上编辑/etc/ssh/sshd_config文件,将PermitRootLogin prohibit-password改为PermitRootLogin yes。

重新启动ssh。

```
root@ubuntu-chain1:/etc/ssh# service ssh restart
```

编辑sudoers文件,添加Defaults visiblepw一行。

```
root@blockchain:/etc# vi sudoers
Defaults        visiblepw
```

2. 在 master 主机上生成密钥对

用ssh-keygen命令生成密钥。

```
root@blockchain:~# ssh-keygen
Generating public/private rsa key pair.
Enter file in which to save the key (/root/.ssh/id_rsa):    #采用默认值
Enter passphrase (empty for no passphrase):                 #第一次输入密钥
Enter same passphrase again:                                #第二次输入密钥
Your identification has been saved in /root/.ssh/id_rsa.
Your public key has been saved in /root/.ssh/id_rsa.pub.
The key fingerprint is:
SHA256:Gt8GsK/+ytW7tyhAmD5nekENOQi/V/dfkNp6fv9ZEn4 root@blockchain
The key's randomart image is:
+---[RSA 2048]----+
|  . . .          |
|  .. +       . |
|   .o.+. .    o  |
|   o.o+.. . o .  |
|   ..o+ S  o o . |
|    o.== +   + o |
|    =.o+ + . = E |
|    ...o.. o.o o+|
|     o=o..+o...o=|
+----[SHA256]-----+
```

3. 将密钥复制到远程主机

```
root@blockchain:~# ssh-copy-id root@192.168.47.139
/usr/bin/ssh-copy-id: INFO: Source of key(s) to be installed: "/root/.ssh/id_rsa.pub"
The authenticity of host '192.168.47.139 (192.168.47.139)' can't be established.
ECDSA key fingerprint is SHA256:OWyao27WG5oDyMeuPwYd7eRjXl66wRB6Wp5WJTEsQN8.
Are you sure you want to continue connecting (yes/no)? yes
/usr/bin/ssh-copy-id: INFO: attempting to log in with the new key(s), to filter out any that are already installed
/usr/bin/ssh-copy-id: INFO: 1 key(s) remain to be installed -- if you are prompted now it is to install the new keys
root@192.168.47.139's password:
Number of key(s) added: 1
Now try logging into the machine, with:   "ssh 'root@192.168.47.139'"
and check to make sure that only the key(s) you wanted were added.
```

4. 测试登录

出现如下信息说明登录成功。

```
root@blockchain:/etc# ssh root@192.168.47.139
Welcome to Ubuntu 18.04.2 LTS (GNU/Linux 4.15.0-91-generic x86_64)

 * Documentation:  https://help.ubuntu.com
 * Management:     https://landscape.canonical.com
 * Support:        https://ubuntu.com/advantage

  System information as of Tue Apr  7 12:54:26 UTC 2020

  System load:  0.16              Processes:             184
  Usage of /:   16.6% of 39.12GB  Users logged in:       1
  Memory usage: 5%                IP address for ens32:  192.168.47.139
  Swap usage:   0%

 * Kubernetes 1.18 GA is now available! See https://microk8s.io for docs or
   install it with:

     sudo snap install microk8s --channel=1.18 --classic

 * Multipass 1.1 adds proxy support for developers behind enterprise
   firewalls. Rapid prototyping for cloud operations just got easier.

     https://multipass.run/
102 packages can be updated.
0 updates are security updates.

*** System restart required ***
```

用exit命令退出登录。

```
root@ubuntu-chain1:~# exit
logout
Connection to 192.168.47.139 closed.
```

上面登录成功出现的提示符是"root@ubuntu-chain1:~#"，说明已经登录成功。

5. 添加 ubuntu-chain1 主机

执行如下命令，如果有错误可以重复几次。

```
root@blockchain:~# docker-machine create --driver generic --generic-ip-address=192.168.47.139 chain1
Running pre-create checks...
Creating machine...
(chain0) No SSH key specified. Assuming an existing key at the default location.
Waiting for machine to be running, this may take a few minutes...
Detecting operating system of created instance...
Waiting for SSH to be available...
Detecting the provisioner...
Provisioning with ubuntu(systemd)...
Installing Docker...
Copying certs to the local machine directory...
Copying certs to the remote machine...
Setting Docker configuration on the remote daemon...
Checking connection to Docker...
Docker is up and running!
```

```
    To see how to connect your Docker Client to the Docker Engine running on
this virtual machine, run: docker-machine env chain1
```

下面是命令中几个参数的说明。

-d driver：指定基于什么虚拟化技术的驱动。

--generic-ip-address：指定要安装宿主机的IP，这里是本地IP。也就是说，可以给别的主机装Docker，前提是SSH root用户免交互登录或私钥认证。

--generic-ssh-user：SSH的用户。

--generic-key-key：指定私钥来实现免交互登录。

6. 测试 docker-machine 主机信息列表

执行docker-machine ls命令查看是否有添加到的主机。

```
root@blockchain:~# docker-machine ls
NAME     ACTIVE   DRIVER    STATE     URL                        SWARM   DOCKER    ERRORS
Chain1   -        generic   Running   tcp://192.168.47.139:2376          v19.03.8
```

七、docker-machine 命令的使用

docker-machine有较多的命令，可以用help参数查看帮助信息。下面举几个常用的命令。

1. 查看 docker-machine 主机的环境变量

```
root@blockchain:~# docker-machine env chain1
export DOCKER_TLS_VERIFY="1"
export DOCKER_HOST="tcp://192.168.47.139:2376"
export DOCKER_CERT_PATH="/root/.docker/machine/machines/chain1"
export DOCKER_MACHINE_NAME="chain1"
# Run this command to configure your shell:
# eval $(docker-machine env chain1)
```

根据上面环境变量信息的最后两行显示，运行eval $(docker-machine env chain1)命令可以配置自己的shell，下面执行该命令。

```
[root@blockchain ~]# eval $(docker-machine env chain0)
[root@blockchain ~ [chain1]]#
```

该命令执行后，命令提示符发生变化，在命令提示符后加上"[chain1]"，说明当前执行命令时，是对chain1主机进行的操作，这样很方便识别。这个功能就是前面设置PS1='[\u@\h \W$(__docker_machine_ps1)]\$ '变量的作用。

2. 运行容器 busubox 测试

```
[root@blockchain ~ [chain1]]# docker run -itd busybox
e7caa93d72a69f20eaa6701d971d1ee750be4ae0bd8d1dccd4b875249e3c492d
[root@blockchain ~ [chain1]]# docker ps
CONTAINER ID   IMAGE     COMMAND   CREATED          STATUS    PORTS     NAMES
e7caa93d72a6   busybox   "sh"      10 seconds ago   Up 8 seconds        cranky_bouman
```

从查看的结果上看，在docker-machine主机chain1上运行一个容器，ID是e7caa93d72a6。再回到docker-machine原主机上运行docker ps命令，查看运行的容器信息，可以看到运行容器的ID是e7caa93d72a6，正是同一个容器，说明在docker-machine的master主机上查看的信息确实是docker-machine主机的信息。

```
root@chain0:~# docker ps
CONTAINER ID     IMAGE          COMMAND      CREATED           STATUS
PORTS            NAMES
 e7caa93d72a6    busybox        "sh"         55 seconds ago    Up 53 seconds
cranky_bouman
```

3. 查看 docker-machine 主机的 Docker 版本

```
[root@blockchain ~ [chain1]]# docker-machine version chain1
19.03.8
```

4. 更新 docker-machine 主机的 Docker 版本

```
[root@blockchain ~ [chain1]]# docker-machine  upgrade chain1
Waiting for SSH to be available...
Detecting the provisioner...
Upgrading docker...
Restarting docker...
```

5. 获取 docker-machine 主机地址

```
[root@blockchain ~ [chain1]]# docker-machine ip chain1
192.168.47.139
```

6. 获取 docker-machine 主机状态

```
[root@blockchain ~ [chain1]]# docker-machine status chain1
Running
```

7. 查看主机的配置信息

```
[root@blockchain ~ [chain1]]# docker-machine config chain1
--tlsverify
--tlscacert="/root/.docker/machine/machines/chain1/ca.pem"
--tlscert="/root/.docker/machine/machines/chain1/cert.pem"
--tlskey="/root/.docker/machine/machines/chain1/key.pem"
-H=tcp://192.168.47.139:2376
```

至此，完成了docker-machine的安装，配置和使用，将其中的ubuntu-chain1主机加入了docker-machine集群。该工具方便进行批量docker-machine主机的管理与安装，提高了效率，方便了管理工作。

任务总结

① docker-machine、docker-compose和随后要学习的docker-swarm并称docker的三剑客，是重要的docker工具，本次学习的docker-machine在部署多主机的docker环境起到重要作用，大大提高了部署效率。

② 加入docker-machine集群的主机需要认证，需要将公钥复制到docker-machine客户主机上，但是要用root权限来安装，在ubuntu系统下，默认禁止root用户登录系统，因此需要先配置root允许SSH登录系统并免交互登录或指定私钥登录。

③ 提示符补全的目的是在完成docker-machine客户主机加入之后，在管理过程中提示符会显

示管理某台docker-machine客户主机，这样管理起来比较直观方便。可以省略该步骤，如果随后想禁用该功能，可将$HOME/.bashrc文件中的PS1='[\u@\h \W$(__docker_machine_ps1)]\$ '注释掉即可。

任务扩展

本任务只是将ubuntu-chain1主机添加到了docker-machine集群中，读者尝试将ubuntu-chain0主机添加到docker-machine集群中。

任务二　分析多主机 overlay 网络

任务场景

安安公司决定采用多主机模式承载公司的网络业务，为增强技术人员对docker网络的理解，提升维护能力，要求技术人员全面学习多主机模式下的overlay网络架构，分析overlay网络桥接关系。

任务描述

本任务学习多主机模式下的overlay网络架构，分析overlay网络桥接关系。

任务目标

◎ 能够描述overlay技术的架构
◎ 能够描述vxlan的协议层次和逻辑结构
◎ 能够描述vxlan的协议封装结构
◎ 能够通过配置consul数据库实现overlay网络多主机通信
◎ 能够运用overlay网络创建容器，并测试容器跨主机通信
◎ 能够根据提示分析overlay网络桥接关系，绘制overlay网络桥接关系图

任务实施

一、认识 Overlay 技术

Overlay是一种网络架构上叠加的虚拟化技术模式，是物理网络向云和虚拟化的深度延伸，使云资源池化能力可以摆脱物理网络的重重限制，是实现云网融合的关键。

1. overlay 技术需求

由于基于虚拟化和云计算技术的发展，传统物理网络架构遇到了重大挑战。

（1）虚拟机迁移范围受到网络架构限制。由于虚拟机迁移的网络属性要求，其从一个物理机上迁移到另一个物理机上，要求虚拟机不间断业务，则需要其IP地址、MAC地址等参数保持不变，如此则要求业务网络是一个二层网络，且要求网络本身具备多路径多链路的冗余和可靠性。

（2）虚拟机规模受网络规格限制。在大二层网络环境下，数据流均需要通过明确的网络寻

141

址以保证准确到达目的地,因此网络设备的二层地址表项大小(即MAC地址表)成为决定云计算环境下虚拟机规模的上限。

(3) 网络隔离/分离能力限制。当前的主流网络隔离技术为VLAN(或VPN),在大规模虚拟化环境部署会有问题,VLAN数量在标准定义中只有12个比特单位,即可用的数量为4 000多个,这样的数量级对于公有云或大型虚拟化云计算应用而言微不足道。在此驱动力基础上,逐步演化出overlay的虚拟化网络技术趋势。

2. overlay 技术简介

overlay在网络技术领域是指一种网络架构上叠加的虚拟化技术模式,其大体框架是对基础网络不进行大规模修改的条件下,实现应用在网络上的承载,并能与其他网络业务分离,并且以基于IP的基础网络技术为主。其实这种模式是以对传统技术的优化而形成的。

针对前文提出的三大技术挑战,Overlay在很大程度上提供了全新的解决方案。

① 针对虚机迁移范围受到网络架构限制的解决方式。Overlay是一种封装在IP报文之上的数据格式,因此,这种数据可以通过路由的方式在网络中分发,而路由网络本身并无特殊网络结构限制,具备良性大规模扩展能力,并且对设备本身无特殊要求,以高性能路由转发为佳,且路由网络本身具备很强的故障自愈能力、负载均衡能力。

② 针对虚机规模受网络规格限制的解决方式。虚拟机数据封装在IP数据包中后,对网络只表现为封装后的网络参数,即隧道端点的地址,因此,对于承载网络(特别是接入交换机),MAC地址规格需求极大降低,最低规格也就是几十个(每个端口一台物理服务器的隧道端点MAC)。

③ 针对网络隔离/分离能力限制的解决方式。针对VLAN数量4 000的限制,在overlay技术中引入了类似12比特VLAN ID的用户标识,支持千万级以上的用户标识,并且在overlay中沿袭了云计算"租户"的概念,称为Tenant ID(租户标识),用24比特或64比特表示。针对VLAN技术下网络的TRUANK ALL(VLAN穿透所有设备)的问题,overlay对网络的VLAN配置无要求,可以避免网络本身的无效流量带宽浪费,同时overlay的二层连通基于虚机业务需求创建,在云的环境中全局可控。

3. VXLAN 技术

overlay主要技术标准中,目前主要有VXLAN、NVGRE和STT技术,而VXLAN利用了现有通用的UDP传输,成熟性相对较高,下面主要对VXLAN进行介绍。

VXLAN(Virtual eXtensible Local Area Network,虚拟扩展局域网)是由IETF定义的NVO3(Network Virtualization over Layer 3)标准技术之一,采用L2 over L4(MAC-in-UDP)的报文封装模式,将二层报文用三层协议进行封装,可实现二层网络在三层范围内进行扩展,同时满足数据中心大二层虚拟迁移和多租户的需求。

(1) VXLAN网络模型(见图6-2)

从图6-2中可以发现,VXLAN网络中出现了以下传统数据中心网络中没有的新元素:

① VTEP(VXLAN Tunnel Endpoints,VXLAN隧道端点)。VXLAN网络的边缘设备,是VXLAN隧道的起点和终点,VXLAN报文的相关处理均在这上面进行。总之,它是VXLAN网络中绝对的主角。VTEP既可以是一个独立的网络设备(如华为的CE系列交换机),也可以是虚拟机所在的服务器。

② VNI(VXLAN Network Identifier,VXLAN 网络标识符)。前文提到,以太网数据帧中VLAN只占了12比特的空间,这使得VLAN的隔离能力在数据中心网络中力不从心。而VNI就是专门解决这个问题的。VNI是一种类似于VLAN ID的用户标示,一个VNI代表了一个租户,属于不同VNI的虚拟机之间不能直接进行二层通信。VXLAN报文封装时,给VNI分配了足够的空间

使其可以支持海量租户的隔离。

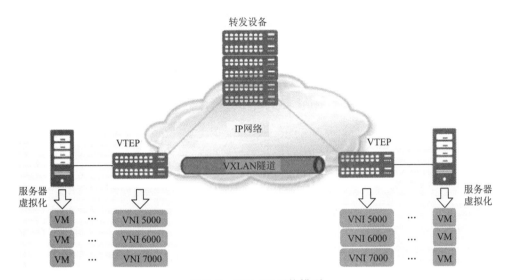

图 6-2　VXLAN 网络模型

③ VXLAN隧道。"隧道"是一个逻辑上的概念，就好像原始报文的起点和终点之间，有一条直通的链路一样。而这个看起来直通的链路就是"隧道"。顾名思义，"VXLAN隧道"用来传输经过VXLAN封装的报文，它是建立在两个VTEP之间的一条虚拟通道。

(2) VXLAN报文格式

VXLAN的报文格式如图6-3所示。下面对主要字段进行解释。

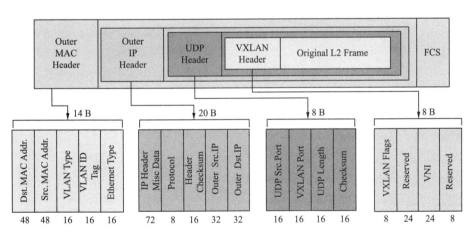

图 6-3　VXLAN 报文格式

`VXLAN Header`

增加VXLAN头（8字节），其中包含24比特的VNI字段，用来定义VXLAN网络中不同的租户。此外，还包含VXLAN Flags（8比特，取值为00001000）和两个保留字段（分别为24比特和8比特）。

`UDP Header`

VXLAN头和原始以太帧一起作为UDP的数据。UDP头中，目的端口号（VXLAN Port）固定为4789，源端口号（UDP Src. Port）是原始以太帧通过哈希算法计算后的值。

`Outer IP Header`

封装外层IP头。其中，源IP地址（Outer Src. IP）为源VM所属VTEP的IP地址，目的IP地址（Outer Dst. IP）为目的VM所属VTEP的IP地址。

`Outer MAC Header`

封装外层以太头。其中，源MAC地址（Src. MAC Addr.）为源VM所属VTEP的MAC地址，目的MAC地址（Dst. MAC Addr.）为到达目的VTEP的路径上下一跳设备的MAC地址。

（3）VXLAN报文转发机制

VXLAN报文转发机制如图6-4所示，通过VXLAN隧道，"二层域"可以突破物理上的界限，实现大二层网络中VM之间的通信。所以，连接在不同VTEP上的VM之间如果有大二层互通的需求，这两个VTEP之间就需要建立VXLAN隧道。换言之，同一大二层域内的VTEP之间都需要建立VXLAN隧道。

例如，假设图6-4中VTEP_1连接的VM、VTEP_2连接的VM及VTEP_3连接的VM之间需要大二层互通，那VTEP_1、VTEP_2和VTEP_3之间就需要两两建立VXLAN隧道，如图6-5所示。

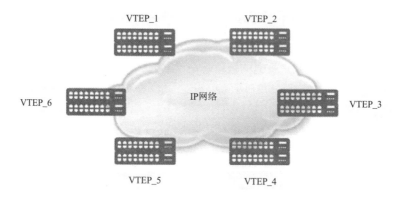

图 6-4 VXLAN 报文转发机制

二、通过 consul 实现的 overlay 实现多主机通信

Docker实现多主机通信，可以通过使用Docker Machine和Docker Swarm来完成Docker跨主机网络的搭建，也可以通过第三方key-value存储服务器搭建Docker实现多主机通信。本任务通过consul实现的overlay实现多主机通信。

1. 架构设计

本任务采用3台宿主机：blockchain、ubuntu-chain0和ubuntu-chain1。其中blockchain作为服务注册主机，安装consul，提供key-value存储，ubuntu-chain0主机和ubuntu-chain1主机作为注册的docker主机。consul实现overlay实现多主机通信架构如图6-6所示。

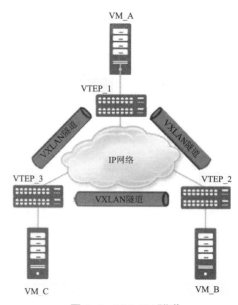

图 6-5 VXLAN 隧道

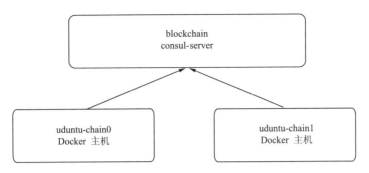

图 6-6　consul 实现 overlay 实现多主机通信架构

2. 运行 consul 服务

consul 服务作为容器提供服务，端口为 8500，下面运行该容器。

```
root@blockchain:~#docker run -d -p 8500:8500  --name consul-server consul agent -client 0.0.0.0 -ui -server -bootstrap
Unable to find image 'consul:latest' locally
latest: Pulling from library/consul
...   （此处省略了部分显示信息）
Digest: sha256:c84237fba5681416f903c22ad5830cbbc26aabe8f9eef466f7ec45b0c70e69f9
Status: Downloaded newer image for consul:latest
b0247238d3028ee2d4b01393ac898012ed5472f91aebf0d82194328dafcc34bf
```

查看运行状态。

```
root@blockchain:~# docker ps -a
CONTAINER ID     IMAGE          COMMAND              CREATED         STATUS
PORTS                  NAMES
b0247238d302     consul         "docker-entrypoint.s…"   26 seconds ago
Up 24 seconds       8300-8302/tcp, 8301-8302/udp, 8600/tcp, 8600/udp, 0.0.0.0:8500->8500/tcp    consul-server
```

Consul 命令参数介绍：

agent：表示启动 Agent 进程。

-server：表示启动 Consul Server 模式。

-client：表示启动 Consul Client 模式。

-bootstrap：表示这个节点是 Server-Leader，每个数据中心只能运行一台服务器。技术角度上讲 Leader 是通过 Raft 算法选举的，但是集群第一次启动时需要一个引导 Leader。在引导群集后，不建议使用此标志。

-ui：表示启动 Web UI 管理器，默认开放端口 8500，所以上面使用 docker 命令把 8500 端口对外开放。

-node：节点的名称，集群中必须是唯一的。

-client：表示 Consul 将绑定客户端接口的地址，0.0.0.0 表示所有地址都可以访问。

-join：表示加入到某一个集群中去。如 -json=192.168.1.23。

3. 通过 web 界面访问管理

在浏览器中输入 http://192.168.47.132:8500/ui/dc1/services 进入管理界面，如图 6-7 所示，

consul已经正常提供服务。

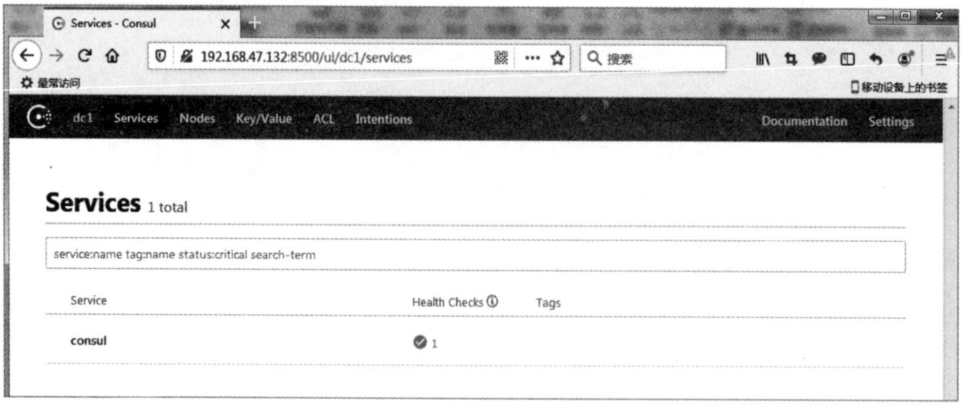

图 6-7 访问 consul 服务界面

4. 查看 docker 主机信息

查看ubuntu-chain0和ubuntu-chain1主机的网络信息，明确主机的IP地址。

（1）查看ubuntu-chain0主机的网络信息

该网卡是ens32，IP地址是192.168.47.137。

```
root@ubuntu-chain0:~# ifconfig
…  （此处省略了部分显示信息）
ens32: flags=4163<UP,BROADCAST,RUNNING,MULTICAST>  mtu 1500
        inet 192.168.47.137  netmask 255.255.255.0  broadcast 192.168.47.255
        inet6 fe80::20c:29ff:fe87:20f6  prefixlen 64  scopeid 0x20<link>
        ether 00:0c:29:87:20:f6  txqueuelen 1000  (Ethernet)
        RX packets 87860  bytes 121839218 (121.8 MB)
        RX errors 0  dropped 0  overruns 0  frame 0
        TX packets 18591  bytes 1260792 (1.2 MB)
        TX errors 0  dropped 0 overruns 0  carrier 0  collisions 0
…  （此处省略了部分显示信息）
```

（2）查看ubuntu-chain1主机的网络信息

该网卡是ens32，IP地址是192.168.47.139。

```
root@ubuntu-chain1:~# ifconfig
…  （此处省略了部分显示信息）
ens32: flags=4163<UP,BROADCAST,RUNNING,MULTICAST>  mtu 1500
        inet 192.168.47.139  netmask 255.255.255.0  broadcast 192.168.47.255
        inet6 fe80::20c:29ff:fe7a:a1cc  prefixlen 64  scopeid 0x20<link>
        ether 00:0c:29:7a:a1:cc  txqueuelen 1000  (Ethernet)
        RX packets 5135  bytes 4879776 (4.8 MB)
        RX errors 0  dropped 0  overruns 0  frame 0
        TX packets 1382  bytes 129871 (129.8 KB)
        TX errors 0  dropped 0 overruns 0  carrier 0  collisions 0
```

…　（此处省略了部分显示信息）

经过查看ubuntu-chain0和ubuntu-chain1主机信息，可知该两台主机属于同一网段。

5. **修改 ubuntu-chain0 和 ubuntu-chain1 主机的 docker 配置文件**

修改ubuntu-chain0和ubuntu-chain1主机的docker配置文件，使ubuntu-chain0和ubuntu-chain1主机启动docker时自动注册到consul服务器中。

在Docker的启动选项里，添加-H tcp://0.0.0.0:2376和--cluster-store选项内容。表示Docker引擎监听所有网络接口上的2376端口，--cluster-store指定了Consul服务，--cluster-advertise表示Docker引擎要注册的网络接口。

（1）修改ubuntu-chain0主机的docker配置文件并重启

先备份 /lib/systemd/system/docker.service文件。

```
root@ubuntu-chain0:~# cp /lib/systemd/system/docker.service /lib/systemd/system/docker.service.bk
```

修改/lib/systemd/system/docker.service文件，如下蓝色显示部分，修改完后保存。

```
[Service]
…　（此处省略了部分显示信息）
ExecStart=/usr/bin/dockerd -H tcp://0.0.0.0:2376 -H fd:// --containerd=/run/containerd/containerd.sock --cluster-store=consul://192.168.47.132:8500 --cluster-advertise=ens32:2376
```

重新加载docker服务。

```
root@ubuntu-chain0:~# systemctl daemon-reload
```

（2）修改ubuntu-chain1主机的docker配置文件并重启

用同样的方法先备份和修改。

```
root@ubuntu-chain1:~# cp /lib/systemd/system/docker.service /lib/systemd/system/docker.service.bk
```

修改/lib/systemd/system/docker.service文件，如下蓝色显示部分，修改完后保存。

```
[Service]
Type=notify
…　（此处省略了部分显示信息）
ExecStart=/usr/bin/dockerd -H tcp://0.0.0.0:2376 -H fd:// --containerd=/run/containerd/containerd.sock --cluster-store=consul://192.168.47.132:8500 --cluster-advertise=ens32:2376
```

重新加载docker服务。

```
root@ubuntu-chain1:~# systemctl daemon-reload
```

6. **查看主机支持的网络**

可以通过docker info 命令查看ubuntu-chain0和ubuntu-chain1主机docker信息，确定支持的网络类型。

（1）查看ubuntu-chain0主机docker信息

```
root@ubuntu-chain0:~# docker info
…　（此处省略了部分显示信息）
```

```
  Plugins:
   Volume: local
   Network: bridge host ipvlan macvlan null overlay
```
……（此处省略了部分显示信息）

（2）查看ubuntu-chain1主机docker信息

```
root@ubuntu-chain1:~# docker info
```
……（此处省略了部分显示信息）
```
  Plugins:
   Volume: local
   Network: bridge host ipvlan macvlan null overlay
```
……（此处省略了部分显示信息）

从以上docker详细信息上看，docker支持的网络有bridge host ipvlan macvlan null overlay。

7. 查看 consul 数据库信息

当docker主机重新启动后会自动向数据库中注册自己的信息。在浏览器中查看kv信息，已经包含了ubuntu-chain0和ubuntu-chain1主机注册信息，如图6-8所示。

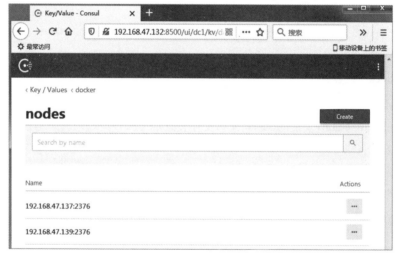

图 6-8　ubuntu-chain0 和 ubuntu-chain1 主机注册成功界面

下面通过创建overlay网络，创建的容器使用overlay网络，然后测试网络的连通性。

8. 创建 overlay 网络

（1）在ubuntu-chain0上创建overlay网络

```
root@ubuntu-chain0:~# docker network create -d overlay my-voernet1
0fac5863f41680e9491e4e1100b3f502dd6cb2af0d6c629e8f5f8c281b24921e
```

这里 -d overlay说明是创建的网络是overlay类型，查看docker网络，可以看到列表中有ID为effc4743a7db，类型为overlay的my-voernet1网络，而且其SCOPE为global，说明具有全局功能，可以与其他主机中的容器通信。

```
root@ubuntu-chain0:~# docker network ls
NETWORK ID          NAME                DRIVER              SCOPE
f32ddb66e4bc        bridge              bridge              local
```

```
46506b451947        host              host              local
effc4743a7db        my-voernet1       overlay           global
365db6c37053        none              null              local
```

在ubuntu-chain1上查看网络信息，从查看的结果上看，列表中有ID为effc4743a7db，类型为overlay的my-voernet1网络，而且其SCOPE为global。即使没有在ubuntu-chain1上创建my-voernet1网络，也可以查看到my-voernet1网络信息，说明ubuntu-chain0主机将信息注册到consul数据库中，ubuntu-chain1主机可以读取共享该信息。

```
root@ubuntu-chain1:~# docker network ls
NETWORK ID          NAME              DRIVER            SCOPE
77d99f963904        bridge            bridge            local
7aced94b59cd        host              host              local
effc4743a7db        my-voernet1       overlay           global
98a755247d45        none              null              local
```

(2) 查看详细信息

用docker network inspect 命令查看my-voernet1网络详细信息。可以看到my-voernet1网络的ID、创建时间、scope、driver等信息，这里注意观察IPAM信息中，网段是10.0.0.0/24，网关是10.0.0.1，也就是其创建隧道中，封装的IP网段是10.0.0.0/24，当创建的容器使用my-voernet1网络时，容器的通信将使用10.0.0.0/24网段地址。

```
root@ubuntu-chain0:~# docker network inspect my-voernet1
[
    {
        "Name": "my-voernet1",
        "Id": "effc4743a7db20d5a2d14d7ac1ee4b06bdbc7babe8220472ae73c3fa050fbe34",
        "Created": "2020-04-11T00:42:26.721419235Z",
        "Scope": "global",
        "Driver": "overlay",
        "EnableIPv6": false,
        "IPAM": {
            "Driver": "default",
            "Options": {},
            "Config": [
                {
                    "Subnet": "10.0.0.0/24",
                    "Gateway": "10.0.0.1"
                }
            ]
        },
        ...   （此处省略了部分显示信息）
    }
]
```

(3) 在consul数据库查看my-voernet1网络信息

打开consul的web页面，可以在docker->network->v1.0->overlay看到my-voernet1网络ID，数据已经记录到数据库中了，如图6-9所示。

图6-9　consul中注册的overlay网络

9. 测试在overlay网络中运行容器

（1）在ubuntu-chain0主机上使用my-voernet1网络运行busybox容器

```
root@ubuntu-chain0:~#docker run -it --name box0 --network my-voernet1 busybox
```

（2）查看box0的IP信息

```
/# ip r
default via 172.18.0.1 dev eth1
10.0.0.0/24 dev eth0 scope link  src 10.0.0.2
172.18.0.0/16 dev eth1 scope link  src 172.18.0.2
```

另外查看本地docker网络信息。

```
root@ubuntu-chain0:~# docker network ls
NETWORK ID          NAME                DRIVER              SCOPE
f32ddb66e4bc        bridge              bridge              local
1fcac3cf1114        docker_gwbridge     bridge              local
46506b451947        host                host                local
effc4743a7db        my-voernet1         overlay             global
365db6c37053        none                null                local
```

可以看到eth0网卡绑定10.0.0.2地址，eth1网卡绑定172.18.0.2地址。实际上eth0网卡绑定的10.0.0.2地址正是overlay类型的my-voernet1网络地址，容器正是将该地址封装到隧道中与外网通信的，而eth1网卡绑定172.18.0.2地址是bridge类型的docker_gwbridge网络，封装的隧道数据将通过该网络进行桥接到物理网卡，从而与外界进行通信。

10. 测试overlay网络容器之间的通信

在ubuntu-chain1主机上使用my-voernet1网络运行busybox容器

```
root@ubuntu-chain1:~# docker run -it --name box1 --network my-voernet1 busybox
/# ip r
```

```
default via 172.18.0.1 dev eth1
10.0.0.0/24 dev eth0 scope link  src 10.0.0.3
172.18.0.0/16 dev eth1 scope link  src 172.18.0.2
```

该容器eth0绑定的IP地址是10.0.0.3，测试与ubuntu-chain0宿主机上box0容器之间的通信。

```
/# ping -c 2 10.0.0.2
PING 10.0.0.2 (10.0.0.2): 56 data bytes
64 bytes from 10.0.0.2: seq=0 ttl=64 time=4.218 ms
64 bytes from 10.0.0.2: seq=1 ttl=64 time=1.106 ms
--- 10.0.0.2 ping statistics ---
2 packets transmitted, 2 packets received, 0% packet loss
round-trip min/avg/max = 0.956/1.858/4.218 ms
```

可以看到ubuntu-chain0宿主机上box0容器正常通信。

11. 测试 overlay 网络的隔离

位于不同主机、overlay类型相同网络的网络的容器之间是可以通信的。下面测试位于不同overlay类型的网络之间的通信。

（1）创建另一个overlay网络

创建一个overlay类型 my-voernet2的网络。

```
root@ubuntu-chain0:~# docker network create -d overlay my-voernet2
9f08337e38f6d59c496720b1a0ef291e9c9ad10e7935f65a2e708453d910fae9
```

（2）分别在ubuntu-chain0和ubuntu-chain1宿主机上创建容器

在ubuntu-chain0宿主机上运行一个box00的容器。

```
root@ubuntu-chain0:~# docker run -it --name box00 --network my-voernet2 busybox
/# ip r
default via 172.18.0.1 dev eth1
10.0.1.0/24 dev eth0 scope link  src 10.0.1.2
172.18.0.0/16 dev eth1 scope link  src 172.18.0.3
```

该容器的IP地址是10.0.1.3。

在ubuntu-chain1宿主机上运行一个box11的容器。

```
root@ubuntu-chain1:~# docker run -it --name box11 --network my-voernet2 busybox
/# ip r
default via 172.18.0.1 dev eth1
10.0.1.0/24 dev eth0 scope link  src 10.0.1.3
172.18.0.0/16 dev eth1 scope link  src 172.18.0.3
```

该容器的IP地址是10.0.1.3。

（3）测试容器box00和box11的通信

```
/# ping -c 2 10.0.1.3
PING 10.0.1.3 (10.0.1.3): 56 data bytes
64 bytes from 10.0.1.3: seq=1 ttl=64 time=1.132 ms
```

```
64 bytes from 10.0.1.3: seq=2 ttl=64 time=1.895 ms
--- 10.0.1.3 ping statistics ---
2 packets transmitted, 2 packets received, 0% packet loss
round-trip min/avg/max = 1.132/1.637/1.895 ms
```

通过测试容器box00和box11是可以相互通信的。

（4）测试my-voernet1和my-voernet2网络上容器之间的连通性

在box00容器上测试到box0和box1的连通性。

```
/# ping -c 3 10.0.0.2
PING 10.0.0.2 (10.0.0.2): 56 data bytes
--- 10.0.0.2 ping statistics ---
3 packets transmitted, 0 packets received, 100% packet loss
/# ping -c 3 10.0.0.3
PING 10.0.0.3 (10.0.0.3): 56 data bytes
--- 10.0.0.3 ping statistics ---
3 packets transmitted, 0 packets received, 100% packet loss
```

从测试的结果上看，my-voernet1和my-voernet2网络上容器之间是不通的，也就是overlay网络之间是隔离的。

三、分析 overlay 网络中 docker_gwbridge 的桥接关系

为了分析overlay网络docker_gwbridge的桥接关系，本任务分别在ubuntu-chain0和ubuntu-chain1主机上创建两个容器。overlay网络通信架构如图6-10所示。

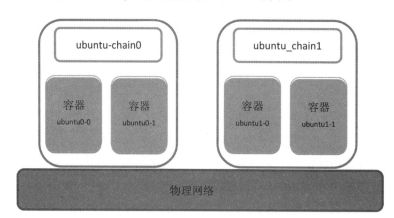

图 6-10　overlay 网络通信架构

1. 在主机上创建容器

（1）在ubuntu-chain0主机上创建ubuntu0-0和ubuntu0-1容器

```
root@ubuntu-chain0:~# docker run -it --name ubuntu0-0 --network my-voernet1 ubuntu
  root@0a60fc1bf9fe:/#
  root@ubuntu-chain0:~# docker run -it --name ubuntu0-1 --network my-voernet2 ubuntu
  root@74363154d3ab:/#
```

(2) 在ubuntu-chain1主机上创建ubuntu1-0和ubuntu1-1容器

```
root@ubuntu-chain1:~# docker run -it --name ubuntu1-0 --network my-voernet1 ubuntu
root@d77c6d99d594:/#
root@ubuntu-chain1:~# docker run -it --name ubuntu1-1 --network my-voernet2 ubuntu
root@58ddf8710d6d:/#
```

2. 在 ubuntu-chain0 和 ubuntu-chain1 宿主机上查看当前运行的容器

(1) 查看ubuntu-chain0宿主机上运行的容器

分别运行的是ubuntu0-0和ubuntu0-1容器。

```
root@ubuntu-chain0:~# docker ps
CONTAINER ID        IMAGE          COMMAND             CREATED           STATUS
PORTS               NAMES
74363154d3ab        ubuntu         "/bin/bash"         2 minutes ago
Up 2 minutes                       ubuntu0-1
0a60fc1bf9fe        ubuntu         "/bin/bash"         3 minutes ago
Up 3 minutes                       ubuntu0-0
```

(2) 查看ubuntu-chain1宿主机上运行的容器

分别运行的是ubuntu1-0和ubuntu1-1容器。

```
root@ubuntu-chain1:~# docker ps
CONTAINER ID        IMAGE          COMMAND             CREATED
STATUS              PORTS          NAMES
58ddf8710d6d        ubuntu         "/bin/bash"         4 minutes ago
Up 4 minutes                       ubuntu1-1
d77c6d99d594        ubuntu         "/bin/bash"         5 minutes ago
Up 4 minutes                       ubuntu1-0
```

3. 查看 ubuntu-chain0 和 ubuntu-chain1 宿主机上网桥信息

(1) 在ubuntu-chain0宿主机上查看网桥信息

用 brctl show命令在ubuntu-chain0查看docker_gwbridge网桥上有两个接口，分别是veth5bc845e和vethdbf63af。

```
root@ubuntu-chain0:~# brctl show
bridge name     bridge id              STP enabled     interfaces
docker0         8000.0242223a43a8      no
docker_gwbridge 8000.0242777eb12e      no              veth5bc845e
                                                       vethdbf63af
```

(2) 在ubuntu-chain1宿主机上查看网桥信息

用 brctl show命令在ubuntu-chain1可以查看到docker_gwbridge网桥上有两个接口，分别是veth5bc845e和vethdbf63af。

```
root@ubuntu-chain1:~# brctl show
bridge name     bridge id              STP enabled     interfaces
docker0         8000.02422177e7b0      no
docker_gwbridge 8000.0242c5fdb581      no              veth4ece85f
```

vethc03769c

4. 查看 ubuntu-chain0 和 ubuntu-chain1 宿主机上网络接口信息

（1）在ubuntu-chain0宿主机上查看网络接口信息

该命令查看到有veth5bc845e和vethdbf63af接口。这两个接口实际就是docker_gwbridge网桥连接到ubuntu0-0和ubuntu0-1容器eth1的网络接口。下面进一步分析。

```
root@ubuntu-chain0:~# ifconfig
…（此处省略了部分显示信息）
docker_gwbridge: flags=4163<UP,BROADCAST,RUNNING,MULTICAST>  mtu 1500
        inet 172.18.0.1  netmask 255.255.0.0  broadcast 172.18.255.255
        inet6 fe80::42:77ff:fe7e:b12e  prefixlen 64  scopeid 0x20<link>
        ether 02:42:77:7e:b1:2e  txqueuelen 0  (Ethernet)
        RX packets 0  bytes 0 (0.0 B)
        RX errors 0  dropped 0  overruns 0  frame 0
        TX packets 20  bytes 1576 (1.5 KB)
        TX errors 0  dropped 0 overruns 0  carrier 0  collisions 0
…（此处省略了部分显示信息）
veth5bc845e: flags=4163<UP,BROADCAST,RUNNING,MULTICAST>  mtu 1500
        inet6 fe80::4cf6:2fff:feea:8199  prefixlen 64  scopeid 0x20<link>
        ether 4e:f6:2f:ea:81:99  txqueuelen 0  (Ethernet)
        RX packets 0  bytes 0 (0.0 B)
        RX errors 0  dropped 0  overruns 0  frame 0
        TX packets 15  bytes 1186 (1.1 KB)
        TX errors 0  dropped 0 overruns 0  carrier 0  collisions 0
vethdbf63af: flags=4163<UP,BROADCAST,RUNNING,MULTICAST>  mtu 1500
        inet6 fe80::f8ea:c3ff:feb5:e490  prefixlen 64  scopeid 0x20<link>
        ether fa:ea:c3:b5:e4:90  txqueuelen 0  (Ethernet)
        RX packets 0  bytes 0 (0.0 B)
        RX errors 0  dropped 0  overruns 0  frame 0
        TX packets 12  bytes 936 (936.0 B)
        TX errors 0  dropped 0 overruns 0  carrier 0  collisions 0
```

（2）在ubuntu-chain1宿主机上查看网络接口信息

同样可以看到veth4ece85f和vethc03769c接口。

```
root@ubuntu-chain1:~# ifconfig
…（此处省略了部分显示信息）
docker_gwbridge: flags=4163<UP,BROADCAST,RUNNING,MULTICAST>  mtu 1500
        inet 172.18.0.1  netmask 255.255.0.0  broadcast 172.18.255.255
        inet6 fe80::42:c5ff:fefd:b581  prefixlen 64  scopeid 0x20<link>
        ether 02:42:c5:fd:b5:81  txqueuelen 0  (Ethernet)
        RX packets 0  bytes 0 (0.0 B)
        RX errors 0  dropped 0  overruns 0  frame 0
        TX packets 18  bytes 1396 (1.3 KB)
        TX errors 0  dropped 0 overruns 0  carrier 0  collisions 0
…（此处省略了部分显示信息）
```

```
veth4ece85f: flags=4163<UP,BROADCAST,RUNNING,MULTICAST>  mtu 1500
        inet6 fe80::a801:99ff:fe13:ab02  prefixlen 64  scopeid 0x20<link>
        ether aa:01:99:13:ab:02  txqueuelen 0  (Ethernet)
        RX packets 0  bytes 0 (0.0 B)
        RX errors 0  dropped 0  overruns 0  frame 0
        TX packets 13  bytes 1006 (1.0 KB)
        TX errors 0  dropped 0 overruns 0  carrier 0  collisions 0
vethc03769c: flags=4163<UP,BROADCAST,RUNNING,MULTICAST>  mtu 1500
        inet6 fe80::c46e:5aff:fe5b:23f4  prefixlen 64  scopeid 0x20<link>
        ether c6:6e:5a:5b:23:f4  txqueuelen 0  (Ethernet)
        RX packets 0  bytes 0 (0.0 B)
        RX errors 0  dropped 0  overruns 0  frame 0
        TX packets 15  bytes 1186 (1.1 KB)
        TX errors 0  dropped 0 overruns 0  carrier 0  collisions 0
```

5. 查看 ubuntu-chain0 和 ubuntu-chain1 宿主机上网络接口序号

用ethtool命令分别在ubuntu-chain0和ubuntu-chain1宿主机上查看veth接口的对端序号。

（1）查看ubuntu-chain0宿主机上veth接口对端序号

在ubuntu-chain0宿主机上用ethtool命令分别查看veth5bc845e和vethdbf63af接口对端序号。

```
root@ubuntu-chain0:~# ethtool -S veth5bc845e
NIC statistics:
     peer_ifindex: 53
root@ubuntu-chain0:~# ethtool -S vethdbf63af
NIC statistics:
     peer_ifindex: 58
```

可以看到veth5bc845e和vethdbf63af接口对端接口序号分别是53和58。

（2）查看ubuntu-chain1宿主机上veth接口对端序号

```
root@ubuntu-chain1:~# ethtool -S veth4ece85f
NIC statistics:
     peer_ifindex: 32
root@ubuntu-chain1:~# ethtool -S vethc03769c
NIC statistics:
     peer_ifindex: 27
```

可以看到veth4ece85f和vethc03769c接口对端接口序号分别是32和27。

6. 查看 ubuntu-chain0 和 ubuntu-chain1 宿主机上容器的接口信息

用ip a命令查看容器接口信息。

（1）查看ubuntu-chain0宿主机上容器的接口信息

查看ubuntu0-0容器上的接口。

```
root@0a60fc1bf9fe:/# ip a
...  （此处省略了部分显示信息）
51: eth0@if52: <BROADCAST,MULTICAST,UP,LOWER_UP> mtu 1450 qdisc noqueue state UP group default
    link/ether 02:42:0a:00:00:02 brd ff:ff:ff:ff:ff:ff link-netnsid 0
```

```
        inet 10.0.0.2/24 brd 10.0.0.255 scope global eth0
           valid_lft forever preferred_lft forever
    53: eth1@if54: <BROADCAST,MULTICAST,UP,LOWER_UP> mtu 1500 qdisc noqueue
state UP group default
        link/ether 02:42:ac:12:00:02 brd ff:ff:ff:ff:ff:ff link-netnsid 1
        inet 172.18.0.2/16 brd 172.18.255.255 scope global eth1
           valid_lft forever preferred_lft forever
```

在ubuntu0-0容器中用ip a命令查看eth0@if52和eth1@if54接口信息，即eth0和eth1接口的MAC地址、IP地址，其序号分别是51和53，这里可以看出，其中53对应的接口就是veth5bc845e，说明ubuntu0-0容器的eth1接口与docker_gwbridge网桥的veth5bc845e接口相连接，至此可以清楚看出这是一对veth pair。

查看ubuntu0-1容器上的接口。

```
root@74363154d3ab:/# ip a
    ...    （此处省略了部分显示信息）
    56: eth0@if57: <BROADCAST,MULTICAST,UP,LOWER_UP> mtu 1450 qdisc noqueue
state UP group default
        link/ether 02:42:0a:00:01:02 brd ff:ff:ff:ff:ff:ff link-netnsid 0
        inet 10.0.1.2/24 brd 10.0.1.255 scope global eth0
           valid_lft forever preferred_lft forever
    58: eth1@if59: <BROADCAST,MULTICAST,UP,LOWER_UP> mtu 1500 qdisc noqueue
state UP group default
        link/ether 02:42:ac:12:00:03 brd ff:ff:ff:ff:ff:ff link-netnsid 1
        inet 172.18.0.3/16 brd 172.18.255.255 scope global eth1
           valid_lft forever preferred_lft forever
```

在ubuntu0-1容器中用ip a命令查看eth1@if57和eth1@if59接口信息，即对应eth0和eth1接口。其序号分别是56和58。可以看出，其中58对应的接口就是vethdbf63af，说明ubuntu0-1容器的eth1接口与docker_gwbridge网桥的vethdbf63af接口相连接，可以看出这是一对veth pair。

（2）查看ubuntu-chain1宿主机上容器的接口信息

查看ubuntu1-0容器上的接口。

```
root@d77c6d99d594:/# ip a
    ...    （此处省略了部分显示信息）
    25: eth0@if26: <BROADCAST,MULTICAST,UP,LOWER_UP> mtu 1450 qdisc noqueue
state UP group default
        link/ether 02:42:0a:00:00:03 brd ff:ff:ff:ff:ff:ff link-netnsid 0
        inet 10.0.0.3/24 brd 10.0.0.255 scope global eth0
           valid_lft forever preferred_lft forever
    27: eth1@if28: <BROADCAST,MULTICAST,UP,LOWER_UP> mtu 1500 qdisc noqueue
state UP group default
        link/ether 02:42:ac:12:00:02 brd ff:ff:ff:ff:ff:ff link-netnsid 1
        inet 172.18.0.2/16 brd 172.18.255.255 scope global eth1
           valid_lft forever preferred_lft forever
```

在ubuntu1-0容器中用ip a命令查看eth1@if26和eth1@if28接口信息，即eth0和eth1接口的

MAC 地址、IP 地址，其序号分别是25和27，这里可以看出，其中27对应的接口就是vethc03769c，说明ubuntu1-0容器的eth1接口与docker_gwbridge网桥的vethc03769c接口相连接，可以看出这是一对veth pair。

查看ubuntu1-1容器上的接口。

```
root@58ddf8710d6d:/# ip a
…   （此处省略了部分显示信息）
30: eth0@if31: <BROADCAST,MULTICAST,UP,LOWER_UP> mtu 1450 qdisc noqueue
state UP group default
    link/ether 02:42:0a:00:01:03 brd ff:ff:ff:ff:ff:ff link-netnsid 0
    inet 10.0.1.3/24 brd 10.0.1.255 scope global eth0
       valid_lft forever preferred_lft forever
32: eth1@if33: <BROADCAST,MULTICAST,UP,LOWER_UP> mtu 1500 qdisc noqueue
state UP group default
    link/ether 02:42:ac:12:00:03 brd ff:ff:ff:ff:ff:ff link-netnsid 1
    inet 172.18.0.3/16 brd 172.18.255.255 scope global eth1
       valid_lft forever preferred_lft forever
```

在ubuntu1-1容器中用ip a命令查看eth1@if31和eth1@if33接口信息，即对应eth0和eth1接口。其序号分别是30和32。可以看出，其中58对应的接口就是veth4ece85f，说明ubuntu1-1容器的eth1接口与docker_gwbridge网桥的veth4ece85f接口相连接，可以看出这是一对veth pair。

7. 通过查看 docker_gwbridge 网桥进一步验证

（1）查看ubuntu-chain0宿主机上docker_gwbridge网桥详细信息

```
root@ubuntu-chain0:~# docker network  inspect docker_gwbridge
[
    {
        "Name": "docker_gwbridge",
        "Id": "1fcac3cf111454c674a64edbd5593704f147f377b3000dcd3419ae2165235c4b",
        "Created": "2020-04-11T01:13:29.047939769Z",
        "Scope": "local",
        "Driver": "bridge",
        "EnableIPv6": false,
        "IPAM": {
            "Driver": "default",
            "Options": null,
            "Config": [
                {
                    "Subnet": "172.18.0.0/16",
                    "Gateway": "172.18.0.1"
                }
            ]
        },
        "Internal": false,
        "Attachable": false,
```

```
            "Ingress": false,
            "ConfigFrom": {
                "Network": ""
            },
            "ConfigOnly": false,
            "Containers": {
                "0a60fc1bf9fe400b6f07a21eae730fb4d317a7a4105c3680bc974780f2ee3131": {
                    "Name": "gateway_cb91126245e2",
                    "EndpointID": "4481130a8d57cc6c450c5c0501033e5861e8bbe3744099e83b33728b8a3725c1",
                    "MacAddress": "02:42:ac:12:00:02",
                    "IPv4Address": "172.18.0.2/16",
                    "IPv6Address": ""
                },
                "74363154d3aba5eb2a7d32994646107a5116ad54f2e105e6ce957dab6590af1b": {
                    "Name": "gateway_0e7dbba65111",
                    "EndpointID": "1fb8fbd5340d4dacd3e7715a34b7c5294a4111cbb83c158c974a1d4845f45589",
                    "MacAddress": "02:42:ac:12:00:03",
                    "IPv4Address": "172.18.0.3/16",
                    "IPv6Address": ""
                }
            },
            "Options": {
                "com.docker.network.bridge.enable_icc": "false",
                "com.docker.network.bridge.enable_ip_masquerade": "true",
                "com.docker.network.bridge.name": "docker_gwbridge"
            },
            "Labels": {}
    }
]
```

以上用docker network inspect docker_gwbridge查看ubuntu-chain0宿主机上docker_gwbridge网桥详细信息，可看到在docker_gwbridge网桥上运行一个ID是0a60fc1bf9fe开头的容器，该容器的网卡ID是4481130a8d57开头的字串，IP地址是172.18.0.2/16，物理地址是02:42:ac:12:00:03，该容器就是ubuntu0-0容器。另一个ID是74363154d3a开头的容器正是ubuntu0-1的信息。

（2）查看ubuntu-chain1宿主机上docker_gwbridge网桥详细信息

```
root@ubuntu-chain1:~# docker network  inspect docker_gwbridge
[
    {
        "Name": "docker_gwbridge",
        "Id": "2e1184b8299d87ea1c6aaab91dd4bff76ca9b1f510cc3b3c901a1aa04387bbe3",
```

```
            "Created": "2020-04-11T01:47:30.244599541Z",
            "Scope": "local",
            "Driver": "bridge",
            "EnableIPv6": false,
            "IPAM": {
                "Driver": "default",
                "Options": null,
                "Config": [
                    {
                        "Subnet": "172.18.0.0/16",
                        "Gateway": "172.18.0.1"
                    }
                ]
            },
            "Internal": false,
            "Attachable": false,
            "Ingress": false,
            "ConfigFrom": {
                "Network": ""
            },
            "ConfigOnly": false,
            "Containers": {
                "58ddf8710d6d6dc5c8f0c728fef332381575a4fdb6d90127fe8b19932
5f90043": {
                    "Name": "gateway_188fddf33754",
                    "EndpointID": "220f3c5da1847436246f5838e70e9ab7c21951989e
2569a59dbff19410914502",
                    "MacAddress": "02:42:ac:12:00:03",
                    "IPv4Address": "172.18.0.3/16",
                    "IPv6Address": ""
                },
                "d77c6d99d5942098f568d2e2b2217b6032d29c0b81f65837fb78eea
6820ab957": {
                    "Name": "gateway_5ca5da0b4eaa",
                    "EndpointID": "90faeed539f1ccd0976f15d3257d2023229fd19
dd429547288f1094fc056f892",
                    "MacAddress": "02:42:ac:12:00:02",
                    "IPv4Address": "172.18.0.2/16",
                    "IPv6Address": ""
                }
            },
            "Options": {
                "com.docker.network.bridge.enable_icc": "false",
                "com.docker.network.bridge.enable_ip_masquerade": "true",
                "com.docker.network.bridge.name": "docker_gwbridge"
```

```
            },
            "Labels": {}
        }
    ]
```

采用同样的方法可以查看ubuntu-chain1宿主机上docker_gwbridge网桥上挂载的ubuntu1-0和ubuntu1-1容器。

8. 绘制 docker_gwbridge 桥接关系图

经过以上的分析，可以看出ubuntu-chain0宿主机上docker_gwbridge网桥通过veth5bc845e与容器ubuntu0-0的eth1接口连接，通过vethdbf63af与容器ubuntu0-1的eth1接口连接。ubuntu-chain1宿主机上docker_gwbridge网桥通过vethc03769c与容器ubuntu1-0的eth1接口连接，通过veth4ece85f与容器ubuntu1-1的eth1接口连接。docker_gwbridge桥接关系如图6-11所示。

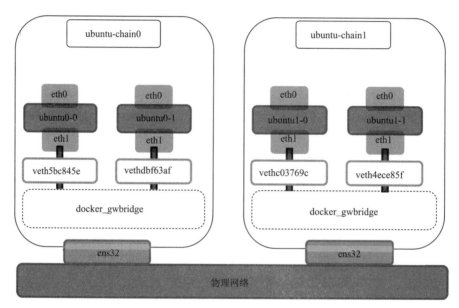

图 6-11　docker_gwbridge 桥接关系

四、分析 overlay 网络中 my-voernet1 和 my-voernet2 的桥接关系

1. 查看 ubuntu-chain0 宿主机上的 namespace

① 创建虚连接。为了查看namespace信息，这里需要创建虚连接。将/var/run/docker/netns 连接到/var/run/netns。

```
root@ubuntu-chain0:~# ln -s /var/run/docker/netns /var/run/netns
```

② 查看/var/run/netns内容。该命令显示的1-9f08337e38和1-effc4743a7就是在ubuntu-chain0宿主机上的两个namespace。

```
root@ubuntu-chain0:~# ls /var/run/netns
1-9f08337e38  1-effc4743a7  bae5cc9ccc71  f1f7750e675c
```

③ 查看1-effc4743a7的桥接信息。在1-effc4743a7命名空间中查看桥接信息。

```
root@ubuntu-chain0:~# ip netns exec 1-effc4743a7 brctl show
```

bridge name	bridge id	STP enabled	interfaces
br0	8000.0e3346d74af4	no	veth0
			vxlan0

这里有一个br0网桥，该桥上面有两个接口，分别是veth0和vxlan0。

④ 查看1-9f08337e38的桥接信息。在1-9f08337e38命名空间中查看桥接信息。

```
root@ubuntu-chain0:~# ip netns exec 1-9f08337e38 brctl show
```

bridge name	bridge id	STP enabled	interfaces
br0	8000.aa7554d4bc6d	no	veth0
			vxlan0

这里同样有一个br0网桥，该桥上面有两个接口，分别是veth0和vxlan0。

⑤ 查看1-effc4743a7的ip信息，在1-effc4743a7命名空间中查看网络信息。

```
root@ubuntu-chain0:~# ip netns exec 1-effc4743a7 ip -d l show
1: lo: <LOOPBACK,UP,LOWER_UP> mtu 65536 qdisc noqueue state UNKNOWN mode DEFAULT group default qlen 1000
    link/loopback 00:00:00:00:00:00 brd 00:00:00:00:00:00 promiscuity 0
addrgenmode eui64 numtxqueues 1 numrxqueues 1 gso_max_size 65536 gso_max_segs 65535
2: br0: <BROADCAST,MULTICAST,UP,LOWER_UP> mtu 1450 qdisc noqueue state UP mode DEFAULT group default
    link/ether 0e:33:46:d7:4a:f4 brd ff:ff:ff:ff:ff:ff promiscuity 0
    ...  （此处省略了部分显示信息）
6: vxlan0@if6: <BROADCAST,MULTICAST,UP,LOWER_UP> mtu 1450 qdisc noqueue master br0 state UNKNOWN mode DEFAULT group default
    link/ether a6:f6:a8:bd:9d:55 brd ff:ff:ff:ff:ff:ff link-netnsid 0 promiscuity 1
    vxlan id 256 srcport 0 0 dstport 4789 proxy l2miss l3miss ttl inherit ageing 300 udpcsum noudp6zerocsumtx noudp6zerocsumrx
    ...  （此处省略了部分显示信息）
8: veth0@if7: <BROADCAST,MULTICAST,UP,LOWER_UP> mtu 1450 qdisc noqueue master br0 state UP mode DEFAULT group default
    link/ether 0e:33:46:d7:4a:f4 brd ff:ff:ff:ff:ff:ff link-netnsid 1 promiscuity 1
    veth
    ...  （此处省略了部分显示信息）
```

该命令查看到共有1: lo、2: br0、6: vxlan0@if6、8: veth0@if7几个接口，可以看到接口的详细配置参数，这里vxlan0@if6接口的序号是6，是挂载在br0上的（蓝色显示），veth0@if7的接口序号是8，是挂载在br0上的（蓝色显示）。对于6: vxlan0@if6接口，可以看到该接口的vxlan ID是256，文中进行了蓝色显示，该接口实际是隧道vetp的端点，与同一网络的另一端vetp端点形成隧道。

⑥ 查看1-9f08337e38的ip信息，在1-9f08337e38命名空间中查看网络信息。

```
root@ubuntu-chain0:~# ip netns exec 1-9f08337e38 ip -d l show
1: lo: <LOOPBACK,UP,LOWER_UP> mtu 65536 qdisc noqueue state UNKNOWN mode DEFAULT group default qlen 1000
```

```
        link/loopback 00:00:00:00:00:00 brd 00:00:00:00:00:00 promiscuity 0
addrgenmode eui64 numtxqueues 1 numrxqueues 1 gso_max_size 65536 gso_max_
segs 65535
    2: br0: <BROADCAST,MULTICAST,UP,LOWER_UP> mtu 1450 qdisc noqueue state
UP mode DEFAULT group default
        link/ether aa:75:54:d4:bc:6d brd ff:ff:ff:ff:ff:ff promiscuity 0
        …（此处省略了部分显示信息）
    11: vxlan0@if11: <BROADCAST,MULTICAST,UP,LOWER_UP> mtu 1450 qdisc noqueue
master br0 state UNKNOWN mode DEFAULT group default
        link/ether ae:ad:f6:d6:23:4c brd ff:ff:ff:ff:ff:ff link-netnsid 0
promiscuity 1
        vxlan id 257 srcport 0 0 dstport 4789 proxy l2miss l3miss ttl inherit
ageing 300 udpcsum noudp6zerocsumtx noudp6zerocsumrx
        …（此处省略了部分显示信息）
    13: veth0@if12: <BROADCAST,MULTICAST,UP,LOWER_UP> mtu 1450 qdisc noqueue
master br0 state UP mode DEFAULT group default
        link/ether aa:75:54:d4:bc:6d brd ff:ff:ff:ff:ff:ff link-netnsid 1
promiscuity 1
        veth
        …（此处省略了部分显示信息）
```

该命令查看到共有1: lo、2: br0、11: vxlan0@if11、13: veth0@if12几个接口，可以看到接口的详细配置参数，这里vxlan0@if11接口的序号是11，是挂载在br0上的（蓝色显示），veth0@if12的接口序号是13，是挂载在br0上的（蓝色显示）。对于11: vxlan0@if11接口，可以看到该接口的vxlan ID是257，文中进行了蓝色显示，该接口实际是隧道vetp的端点，与同一网络的另一端vetp端点形成隧道。

2. 查看和分析接口之间的连接关系

① 查看1-effc4743a7的veth0接口对端连接序号。

```
root@ubuntu-chain0:~#  ip netns exec 1-effc4743a7 ethtool -S veth0
NIC statistics:
     peer_ifindex: 7
```

② 查看1-9f08337e38的veth0接口对端连接序号。

```
root@ubuntu-chain0:~# ip netns exec 1-9f08337e38 ethtool -S veth0
NIC statistics:
     peer_ifindex: 12
```

③ 查看容器ubuntu0-0的eth0的对端连接序号。

```
root@0a60fc1bf9fe:/# ethtool -S eth0
NIC statistics:
     peer_ifindex: 8
```

命令显示容器ubuntu0-0的eth0的对端连接序号是8。

④ 查看容器ubuntu0-1的eth0的对端连接序号。

```
root@74363154d3ab:/# ethtool -S eth0
```

```
    NIC statistics:
         peer_ifindex: 13
```

命令显示容器ubuntu0-1的eth0的对端连接序号是13。

⑤ 查看容器ubuntu0-0的eth0的序号。

```
root@0a60fc1bf9fe:/# ip a
…（此处省略了部分显示信息）
7: eth0@if8: <BROADCAST,MULTICAST,UP,LOWER_UP> mtu 1450 qdisc noqueue state UP group default
    link/ether 02:42:0a:00:00:02 brd ff:ff:ff:ff:ff:ff link-netnsid 0
    inet 10.0.0.2/24 brd 10.0.0.255 scope global eth0
       valid_lft forever preferred_lft forever
…（此处省略了部分显示信息）
```

⑥ 查看容器ubuntu0-1的eth0的序号。

```
root@74363154d3ab:/# ip a
…（此处省略了部分显示信息）
12: eth0@if13: <BROADCAST,MULTICAST,UP,LOWER_UP> mtu 1450 qdisc noqueue state UP group default
    link/ether 02:42:0a:00:01:02 brd ff:ff:ff:ff:ff:ff link-netnsid 0
    inet 10.0.1.2/24 brd 10.0.1.255 scope global eth0
       valid_lft forever preferred_lft forever
…（此处省略了部分显示信息）
```

至此，这种虚拟的连接关系已经很清楚了，容器ubuntu0-0的eth0接口与1-effc4743a7中br0的veth0接口是一对veth pair；容器ubuntu0-1的eth0接口与1-9f08337e38中br0的veth0接口是一对veth pair。从而实现了容器与overlay网络my-voernet1和my-voernet2的连接。下面进一步分析。

3. 查看my-voernet1详细信息

```
root@ubuntu-chain0:~# docker network inspect my-voernet1
[
    {
        "Name": "my-voernet1",
        "Id": "effc4743a7db20d5a2d14d7ac1ee4b06bdbc7babe8220472ae73c3fa050fbe34",
        …（此处省略了部分显示信息）
        "IPAM": {
            "Driver": "default",
            "Options": {},
            "Config": [
                {
                    "Subnet": "10.0.0.0/24",
                    "Gateway": "10.0.0.1"
                }
            ]
        },
```

```
            … （此处省略了部分显示信息）
            "Containers": {
                "0a60fc1bf9fe400b6f07a21eae730fb4d317a7a4105c3680bc974780f2ee3131": {
                    "Name": "ubuntu0-0",
                    "EndpointID": "a379bf85cc5ee909450199318c094816b0fcdae0a580418cb713e2bfcc549bf4",
                    "MacAddress": "02:42:0a:00:00:02",
                    "IPv4Address": "10.0.0.2/24",
                    "IPv6Address": ""
                },
                "ep-2125a55280b61aa020edb044fb1010cb4c68af9c605c6aba27e42a62bee8e753": {
                    "Name": "ubuntu1-0",
                    "EndpointID": "2125a55280b61aa020edb044fb1010cb4c68af9c605c6aba27e42a62bee8e753",
                    "MacAddress": "02:42:0a:00:00:03",
                    "IPv4Address": "10.0.0.3/24",
                    "IPv6Address": ""
                }
            },
            "Options": {},
            "Labels": {}
        }
    ]
```

从显示的信息上看，在my-voernet1网络上挂载有ubuntu0-0和ubuntu1-0容器，文中已蓝色显示。

4. 查看my-voernet2详细信息

```
root@ubuntu-chain0:~# docker network inspect my-voernet2
[
    {
        "Name": "my-voernet2",
        "Id": "9f08337e38f6d59c496720b1a0ef291e9c9ad10e7935f65a2e708453d910fae9",
        … （此处省略了部分显示信息）
        "IPAM": {
            "Driver": "default",
            "Options": {},
            "Config": [
                {
                    "Subnet": "10.0.1.0/24",
                    "Gateway": "10.0.1.1"
                }
            ]
        },
```

```
...     （此处省略了部分显示信息）
        "Containers": {
            "74363154d3aba5eb2a7d32994646107a5116ad54f2e105e6ce957dab
6590af1b": {
                "Name": "ubuntu0-1",
                "EndpointID": "a6e8ec932d517d1c65da13e497c3921e7fd8951
cf650be63faf0a7a23e1cb06e",
                "MacAddress": "02:42:0a:00:01:02",
                "IPv4Address": "10.0.1.2/24",
                "IPv6Address": ""
            },
            "ep-b5ab8566bf857f8c399ae78d63e8c229340137e2f47a3a624d332c
610a3a9aac": {
                "Name": "ubuntu1-1",
                "EndpointID": "b5ab8566bf857f8c399ae78d63e8c229340137e
2f47a3a624d332c610a3a9aac",
                "MacAddress": "02:42:0a:00:01:03",
                "IPv4Address": "10.0.1.3/24",
                "IPv6Address": ""
            }
        },
        "Options": {},
        "Labels": {}
    }
]
```

从显示的信息上看，在my-voernet2网络上挂载有ubuntu0-1和ubuntu1-1容器，文中已蓝色显示。从分配的IP地址和MAC等信息也能进一步验证这种连接关系。

5. 查看1-effc4743a7 命名空间的路由

```
root@ubuntu-chain0:~# ip netns exec 1-effc4743a7 ip route
10.0.0.0/24 dev br0 proto kernel scope link src 10.0.0.1
```

命令显示10.0.0.0/24 通过dev br0设备到达，网关是 10.0.0.1，就是my-voernet1网络。

6. 查看1-9f08337e38 命名空间的路由

```
root@ubuntu-chain0:~# ip netns exec 1-9f08337e38 ip route
10.0.1.0/24 dev br0 proto kernel scope link src 10.0.1.1
```

命令显示10.0.1.0/24 通过dev br0设备到达，网关是 10.0.1.1，就是my-voernet2网络。

7. 查看1-effc4743a7 vxlan0 封装的 vxlan ID

```
root@ubuntu-chain0:~# ip netns exec 1-effc4743a7 ip -d link show vxlan0
6: vxlan0@if6: <BROADCAST,MULTICAST,UP,LOWER_UP> mtu 1450 qdisc noqueue
master br0 state UNKNOWN mode DEFAULT group default
    link/ether a6:f6:a8:bd:9d:55 brd ff:ff:ff:ff:ff:ff link-netnsid 0 promis
cuity 1
    vxlan id 256 srcport 0 0 dstport 4789 proxy l2miss l3miss ttl inherit
```

```
ageing 300 udpcsum noudp6zerocsumtx noudp6zerocsumrx
```
...（此处省略了部分显示信息）

8. 查看 1-9f08337e38 vxlan0 封装的 vxlan ID

```
root@ubuntu-chain0:~# ip netns exec 1-9f08337e38 ip -d link show vxlan0
11: vxlan0@if11: <BROADCAST,MULTICAST,UP,LOWER_UP> mtu 1450 qdisc noqueue master br0 state UNKNOWN mode DEFAULT group default
    link/ether ae:ad:f6:d6:23:4c brd ff:ff:ff:ff:ff:ff link-netnsid 0 promiscuity 1
    vxlan id 257 srcport 0 0 dstport 4789 proxy l2miss l3miss ttl inherit ageing 300 udpcsum noudp6zerocsumtx noudp6zerocsumrx
```
...（此处省略了部分显示信息）

9. 查看隧道对方节点信息

① 在ubuntu-chain0宿主机上查看1-effc4743a7邻居。

```
root@ubuntu-chain0:~# ip netns exec 1-effc4743a7 ip neigh
10.0.0.3 dev vxlan0 lladdr 02:42:0a:00:00:03 PERMANENT
```

② 在ubuntu-chain1宿主机上查看1-effc4743a7邻居。

```
root@ubuntu-chain1:~# ip netns exec 1-effc4743a7 ip neigh
10.0.0.2 dev vxlan0 lladdr 02:42:0a:00:00:02 PERMANENT
```

vxlan0是VXLAN隧道端点，它是VXLAN网络的设备边缘，用于VXLAN报文封包和解包，包括ARP请求报文和正常的VXLAN数据报文，封装好以后报文通过隧道向另一端的VTEP也就是VXLAN隧道端点发送，另一端的VTEP收到后解开报文。

通信过程如下：ubuntu-chain0宿主机上ubuntu0-0容器 ping 其他容器，通过ubuntu0-0容器的eth0发送出去，并通过路由表得知发往br0。br0相当于虚拟交换机，如果目标主机在同一宿主机，则直接通过br0通信；如果不在则通过vxlan通信，br0收到请求会把请求交给vxlan0处理，这时实际上通过vetp处理。

vxlan中保存有MAC地址表（docker守护进程通过gossip协议在concul数据库中学来的），并通过ubuntu-chain0宿主机上的ens32发送出去。报文到达ubuntu-chain1宿主机，拆包发现是vxlan报文，获取IP，则交给它上面的vxlan设备（同一ID的设备），ubuntu-chain1宿主机上vxlan拆包，交给br0，然后br0根据MAC表完成最后的投递。

以上分析是从ubuntu-chain0宿主机上桥接关系进行分析的，ubuntu-chain1宿主机上桥接关系分析方法完全一样，读者可以自行分析。

10. 画出 ubuntu-chain0 宿主机上桥接关系的逻辑结构图

从以上的分析可以看出，ubuntu-chain0宿主机上运行有ubuntu0-0和ubuntu0-1容器，ubuntu-chain1宿主机上运行有ubuntu1-0和ubuntu1-1容器，网络连接关系已经很清楚，下面画出桥接关系的逻辑结构图，如图6-12所示。

任务总结

① VXLAN采用L2 over L4（MAC-in-UDP）的报文封装模式，将二层报文用三层协议进行封装，可实现二层网络在三层范围内进行扩展，同时满足数据中心大二层虚拟迁移和多租户的

需求。这里涉及两个重要的概念：VTEP和VNI。VTEP是VXLAN网络的边缘设备，是VXLAN隧道的起点和终点，VXLAN报文的相关处理均在这上面进行。VNI是一个24bit的标识，用来标识不同的租户，属于不同VNI的虚拟机是不能通信的。

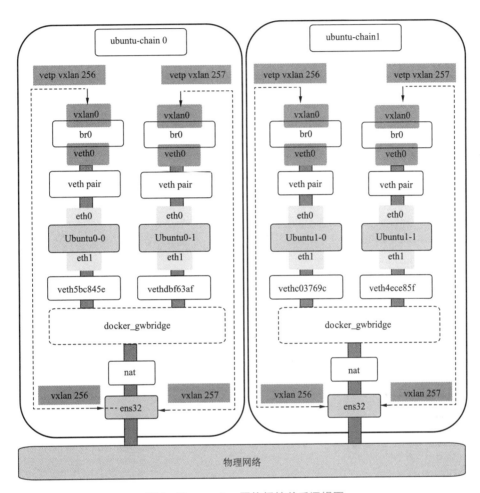

图6-12 overlay网络桥接关系逻辑图

② 本任务采用consul作为key-value数据库，也可以通过其他第三方key-value存储服务器实现多主机通信，如etcd或者Zookeeper等。

③ ip命令功能非常强大，本任务中用到ip命令中的几个命令（ip netns exec命令、ip neigh命令、ip -d l show），读者可以进一步研究命令的格式和用法。

④ overlay网络是学习docker网络中最难的部分，桥接关系复杂，分析命令繁多，如果想深入研究建议多去练习，分析，可以根据网络结构画出桥接关系图，加深理解。

阅读下面附件，认识Zookeeper、etcd、Consul三种工具。

原文：https://technologyconversations.com/2015/09/08/service-discovery-zookeeper-vs-etcd-vs-consul/

摘自：https://www.cnblogs.com/sunsky303/p/11127760.html

使用 docker-swarm 编排网络服务

【项目综述】
　　前期通过 docker-compose 进行编排公司的网络服务，但是存在一个致命的缺点，就是单点故障。随着公司的业务需求，必须解决单点故障的问题，需要选择多主机模式解决网络服务，这需要 docker 的另一个强大工具 docker-swarm。本项目任务是学习 docker-swarm，运用 docker-swarm 部署网络服务。

【项目目标】
- ◎ 能够解释 docker-swarm 关键概念
- ◎ 能够叙述 docker-swarm 架构
- ◎ 能够运用 docker-swarm 部署网络服务

任务一　认识 docker-swarm

任务场景

随着公司的业务需求，公司的网络服务需要部署到多个主机上实现冗余，解决单点故障问题，采用docker-compose编排满足不了需求，需要docker-swarm解决网络服务的编排问题。

任务描述

本任务学习docker-swarm的架构、关键概念和调度策略。

任务目标

- ◎ 能够解释docker-swarm关键概念
- ◎ 能够叙述docker-swarm架构
- ◎ 能够说出docker-swarm调度策略
- ◎ 能够描述docker-swarm的安全机制

任务实施

　　swarm是Docker公司推出的用来管理docker集群的平台，几乎全部用Go语言来完成开发的，代码开源在https://github.com/docker/swarm，它是将一群Docker宿主机变成一个单一的虚拟主机，swarm使用标准的Docker API接口作为其前端的访问入口，换言之，各种形式的Docker Client （compose、docker-py等）均可以直接与swarm通信，甚至Docker本身都可以很容易地与swarm集成，这大大方便了用户将原本基于单节点的系统移植到swarm上，同时swarm内置了对Docker网络插件的支持，用户也很容易地部署跨主机的容器集群服务。

Docker swarm和Docker compose一样，都是 Docker 官方容器编排项目，但不同的是，Docker compose是一个在单个服务器或主机上创建多个容器的工具，而Docker swarm则可以在多个服务器或主机上创建容器集群服务。对于微服务的部署，Docker swarm更加适合。

从Docker 1.12.0 版本开始，Docker swarm已经包含在Docker引擎中（Docker Swarm），并且已经内置了服务发现工具，我们就不需要像之前一样，再配置 Etcd 或者Consul来进行服务发现配置了。

swarm deamon只是一个调度器（scheduler）加路由器（router）。swarm自己不运行容器，它只是接受Docker客户端发来的请求，调度适合的节点来运行容器。这就意味着，即使swarm由于某些原因挂掉了，集群中的节点也会照常运行，swarm重新恢复运行之后，会收集重建集群信息。

一、Docker swarm 主要功能

① Docker engine集成集群管理。使用Docker engine CLI 创建一个Docker engine的swarm模式，在集群中部署应用程序服务。

② 去中心化设计。swarm角色分为manager和worker节点，manager节点故障不影响应用使用。

③ 扩容缩容。可以声明每个服务运行的容器数量，通过添加或删除容器数自动调整期望的状态。

④ 期望状态协调。swarm manager节点不断监视集群状态，并调整当前状态与期望状态之间的差异。

⑤ 多主机网络。可以为服务指定overlay网络。当初始化或更新应用程序时，swarm manager会自动为overlay网络上的容器分配IP地址。

⑥ 服务发现。swarm manager节点为集群中的每个服务分配唯一的DNS记录和负载均衡VIP。可以通过swarm内置的DNS服务器查询集群中每个运行的容器。

⑦ 负载均衡。实现服务副本负载均衡，提供入口访问。

⑧ 安全传输。swarm中的每个节点使用TLS相互验证和加密，确保安全的其他节点通信。

⑨ 滚动更新。升级时，逐步将应用服务更新到节点，如果出现问题，可以将任务回滚到先前版本。

二、认识 Docker swarm 架构

Docker swarm是典型的master-slave结构，通过发现服务来选举manager。manager是中心管理节点，各个node上运行agent接受manager的统一管理，集群会自动通过Raft协议分布式选举出manager节点，无须额外的发现服务支持，避免了单点的瓶颈问题，同时也内置了DNS的负载均衡和对外部负载均衡机制的集成支持。Docker swarm架构如图7-1所示。

三、认识 Docker swarm 的关键概念

1. swarm 节点

swarm是一系列节点的集合，而节点可以是一台裸机或者一台虚拟机。一个节点能扮演一个或者两个角色：manager或者worker，如图7-2所示。

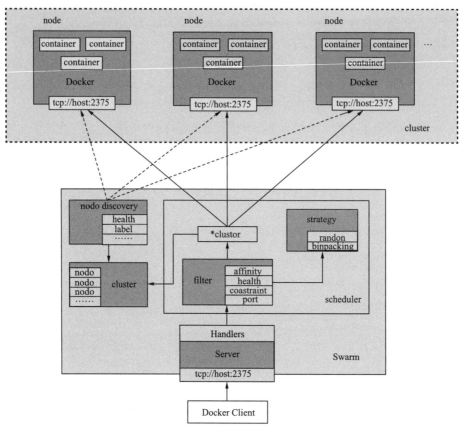

图 7-1　Docker swarm 架构

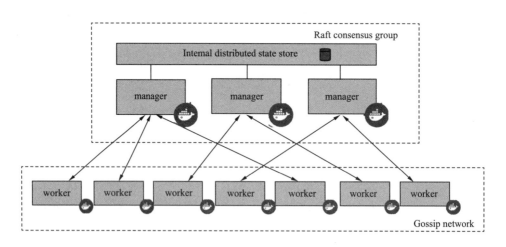

图 7-2　swarm 节点角色

（1）manager节点

执行集群的管理功能，维护集群的状态，选举一个leader节点去执行调度任务。Docker swarm集群需要至少一个manager节点，节点之间使用Raft consensus protocol进行协同工作。通常，第一个启用docker swarm的节点将成为leader，后来加入的都是follower。当前的leader如果挂掉，剩余的节点将重新选举出一个新的leader。每一个manager都有一个完整的当前集群状态的副本，可以保证manager的高可用。

（2）worker节点

接收和执行任务。参与容器集群负载调度，仅用于承载task。worker节点是运行实际应用服务的容器所在的地方。理论上，一个manager节点也能同时成为worker节点，但在生产环境中不建议这样做。worker节点之间，通过control plane进行通信，这种通信使用gossip协议，并且是异步的。

2. service、task 和 stack 关系

在swarm集群中service、task和stack之间的关系，如图7-3所示。

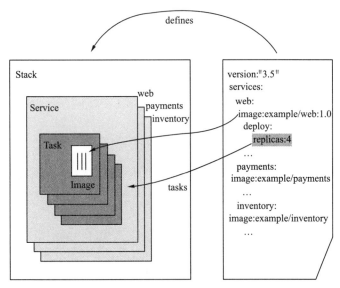

图 7-3　service、task 和 stack 之间关系

（1）service

swarm service是一个抽象的概念，它只是一个对运行在swarm集群上的应用服务，所期望状态的描述。它就像一个描述了下面物品的清单列表一样：

① 服务名称；
② 使用哪个镜像来创建容器；
③ 要运行多少个副本；
④ 服务的容器要连接到哪个网络上；
⑤ 应该映射哪些端口。

典型的服务结构如图7-4所示

（2）task

在Docker swarm中，task是一个部署的最小单元，task与容器是一对一的关系，客户端向manager节点发出请求，manager节点根据指定数量的任务副本分配任务给worker节点，task调度关系如图7-5所示。

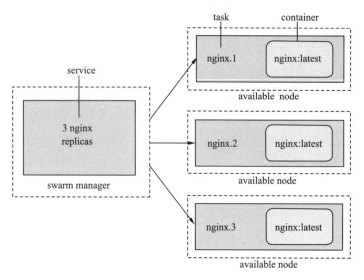

图 7-4 典型的服务结构

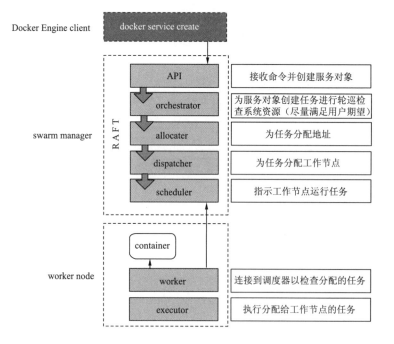

图 7-5 task 调度关系

（3）stack

stack是描述一系列相关services的集合，通过YAML文件定义一个stack，可以一键启动服务集群，可以认为是升级版的compose功能。

（4）docker swarm常用到的命令

docker swarm：集群管理，子命令有init、join、leave、update。（docker swarm --help查看帮助）

docker service：服务创建，子命令有create、inspect、update、remove、tasks。（docker service --help查看帮助）

docker node：节点管理，子命令有accept、promote、demote、inspect、update、tasks、ls、rm。（docker node --help查看帮助）

docker stacks：管理服务集群，子命令有deploy、ls、ps、rm、services。（docker stacks --help查看帮助）

四、swarm 的调度策略

swarm在调度（scheduler）节点（leader节点）运行容器的时候，会根据指定的策略来计算最适合运行容器的节点，目前支持的策略有spread、binpack、random.

1. random

random策略随机选择一个节点来运行容器，一般用作调试用，spread和binpack策略会根据各个节点可用的CPU，RAM以及正在运行容器的数量来计算应该运行容器的节点。

2. spread

在同等条件下，spread策略会选择运行容器最少的那个节点来运行新的容器。

使用spread策略会使得容器会均衡的分布在集群中的各个节点上运行，一旦一个节点挂掉了只会损失少部分的容器。

3. binpack

binpack策略最大化地避免容器碎片化，就是说binpack策略尽可能地把还未使用的节点留给需要更大空间的容器运行，尽可能地把容器运行在一个节点上面，binpack策略会选择运行容器最集中的那台机器来运行新的节点。

五、swarm cluster 模式特性

1. 批量创建服务

建立容器之前先创建一个overlay的网络，用来保证在不同宿主机上的容器网络互通。

2. 强大的集群的容错性

当容器副本中的其中某一个或某几个节点死机后，cluster会根据自己的服务注册发现机制，以及之前设定的值--replicas n，在集群中剩余的空闲节点上，重新拉起容器副本。整个副本迁移的过程无须人工干预，迁移后原本的集群的load balance依旧正常。docker service不仅仅批量启动服务，还在集群中定义了一种状态。cluster会持续检测服务的健康状态并维护集群的高可用性。

3. 服务节点的可扩展性

swarm cluster不仅提供高可用性，同时也提供了节点弹性扩展或缩减的功能。当容器组想动态扩展时，只需通过scale参数即可复制出新的副本出来。如果有需求想在每台节点上都run一个相同的副本，方法只需要在命令中将--replicas n更换成--mode=global即可。

复制服务（--replicas n），将一系列复制任务分发至各节点当中，具体取决于设置状态，如--replicas 3。

全局服务（--mode=global），适用于集群内全部可用节点上的服务任务，如--mode global。

六、swarm 的安全特性

Docker基本支持所有的Linux重要安全技术，同时对其进行封装并赋予合理的默认值，这在

保证了安全的同时也避免了过多的限制，如图7-6所示。

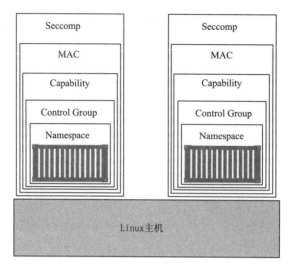

图 7-6　Linux 安全架构

自定义设置某些安全技术会非常复杂，因为这需要用户深入理解安全技术的运作原理，同时还要了解Linux内核的工作机制。希望这些技术在未来能够简化配置的过程，但就现阶段而言，使用 Docker 在对安全技术的封装中提供的默认值是很不错的选择。

Docker平台还引入了大量自有安全技术。swarm模式基于TLS构建，并且配置上极其简单灵活。安全扫描对镜像进行二进制源码级别扫描，并提供已知缺陷的详细报告。

Docker内容信任允许用户对内容进行签名和认证，Docker为这些安全技术设定了合理的默认值，但是用户也可以自行修改配置，或者禁用这些安全技术。

1. Docker swarm 的安全模式

swarm模式是Docker未来的趋势。swarm模式支持用户集群化管理多个Docker主机，还能通过声明式的方式部署应用。

每个swarm都由管理者和工作者节点构成，节点可以是Linux或者Windows。管理者节点构成了集群中的控制层，并负责集群配置以及工作负载的分配。工作者节点就是运行应用代码的容器。

正如所预期的，swarm 模式包括很多开箱即用的安全特性，同时还设置了合理的默认值。这些安全特性包括以下几点：

① 加密节点 ID。
② 基于 TLS 的认证机制。
③ 安全准入令牌。
④ 支持周期性证书自动更新的 CA 配置。
⑤ 加密集群存储（配置 DB）。
⑥ 加密网络。

2. swarm 节点认证关系举例

下面通过例子讲解认证关系。示例中 3 个 Docker 主机分别为mgr1、mgr2、wrk1。

每台主机上都安装 Ubuntu 18.04，其上运行了 Docker 19.03.0-ce。同时还有一个网络负责联通 3 台主机，并且主机之间可以通过名称互相 ping 通。安装完成的认证关系实验环境如图7-7所示。

配置安全的 swarm 集群，在mgr1上运行docker swarm init命令，mgr1被配置为 swarm 集群中的第一个管理节点，也是根 CA 节点。swarm 集群已经被赋予了加密 swarm ID，同时mgr1节点为自己发布了一个客户端认证信息，标明自己是swarm集群管理者。

证书的更新周期默认设置为90天，集群配置数据库也已经配置完成并且处于加密状态。安全令牌也已经成功创建，允许新的管理者和工作者节点加入到 swarm 集群中。

实验环境如图7-8所示。

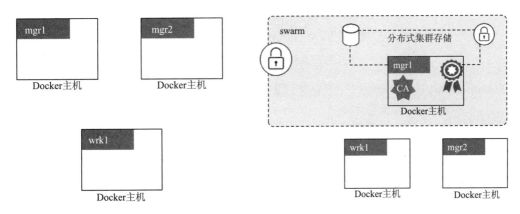

图 7-7 认证关系实验环境　　　　图 7-8 mgr1 认证根节点

现在将mgr2节点加入到集群中，作为额外的管理者节点。

将新的管理者节点加入到 swarm 需要两步。①需要提取加入管理者到集群中所需的令牌；②在mgr2节点上执行 docker swarm join 命令。这样mgr1和mgr2都加入了 swarm，并且都是 swarm 管理者。最新的配置如图7-9所示。

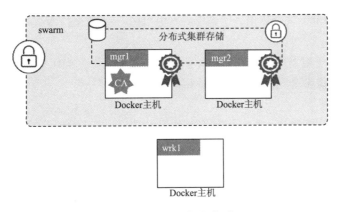

图 7-9 mgr2 加入集群

向 swarm 中加入工作节点也只需两步。①需要获取新工作节点的准入令牌，用docker swarm join-token worker命令获取；②在工作节点上运行 docker swarm join 命令，这样可以创建包含两个管理者和一个工作者的 swarm 集群。管理者配置为高可用（HA），并且复用集群存储。最新的配置如图7-10所示。

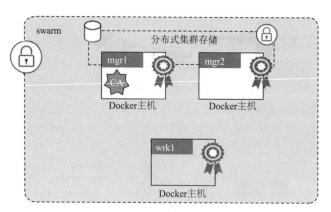

图 7-10　工作节点接入集群

在swarm集群中，需要进行证书认证才能加入到集群中，这样就保证了集群通信的安全。

任务总结

① docker-swarm采用主从分布式结构，由manager主机和worker主机组成，manager主机也可以承担worker主机的功能，manager主机中存储有整个集群的数据库信息。

② docker-swarm集群中stack、service、task之间的关系是：task是docker-swarm部署的最小单元，与容器是一对一的关系；service是由一个或者多个tasks组成，也就是多个容器的副本，提供某一个服务功能；stack是多个service组成，由多个service提供完成某集群的功能。

③ docker-swarm集群通过基于TLS的认证机制，主机节点基本支持所有的Linux重要安全技术，同时对其进行封装并赋予合理的默认值，这在保证了安全的同时也避免了过多的限制。

任务扩展

网上搜索，查阅Linux安全体系，了解Docker中用到Linux安全技术。

1. Namespace

内核命名空间属于容器中非常核心的一部分。该技术能够将操作系统（OS）进行拆分，使一个操作系统看起来像多个互相独立的操作系统一样。

2. Control Group

如果说命名空间用于隔离，那么控制组就是用于限额。在 Docker 的环境中，容器之间是互相隔离的，但却共享 OS 资源，如 CPU、RAM 及硬盘 I/O。CGroup 允许用户设置限制，这样单个容器就不能占用主机全部的 CPU、RAM 或者存储 I/O 资源了。

3. Capability

在在 Docker 的环境中，有时用户需要一种技术，能选择容器运行所需的root用户权限，这个是通过Capability实现。在底层，Linux root用户是由许多能力组成的，其中一部分包括以下几点：

CAP_CHOWN：允许用户修改文件所有权。

CAP_NET_BIND_SERVICE：允许用户将socket绑定到系统端口号。

CAP_SETUID：允许用户提升进程优先级。

CAP_SYS_BOOT：允许用户重启系统。

Docker采用Capability机制来实现用户在以root身份运行容器的同时，还能移除非必需的root能力。如果容器运行只需要root的绑定系统网络端口号的能力，则用户可以在启动容器的同时移除全部root能力，然后再将CAP_NET_BIND_SERVICE能力添加回来。

4. MAC

Docker采用主流Linux MAC技术，如AppArmor及SELinux。

AppArmor（Application Armor的缩写）是强制访问控制（MAC）系统（由Ubuntu Linux，其衍生产品和其他Linux发行版使用），是一个高效和易于使用的Linux系统安全应用程序。AppArmor对操作系统和应用程序所受到的威胁进行从内到外的保护，甚至是未被发现的0day漏洞和未知的应用程序漏洞所导致的攻击。

SELinux（Security-Enhanced Linux）是美国国家安全局（NSA）对于强制访问控制的实现，是Linux历史上最杰出的安全子系统。SELinux是2.6版本的Linux内核中提供的强制访问控制（MAC）系统。对于可用的Linux安全模块来说，SELinux是功能最全面，而且测试最充分的。SELinux 在类型强制服务器中合并了多级安全性或一种可选的多类策略，并采用了基于角色的访问控制概念。

5. Seccomp

Docker使用过滤模式下的Seccomp来限制容器对宿主机内核发起的系统调用。

按照Docker的安全理念，每个新容器都会设置默认的Seccomp配置，文件中设置了合理的默认值。这样做是为了在不影响应用兼容性的前提下提供适度的安全保障。

用户同样可以自定义Seccomp配置，同时也可以通过向Docker传递指定参数，使Docker 启动时不设置任何Seccomp配置。

任务二　利用 docker-swarm 创建 Nginx 集群

任务场景

随着公司的业务需求，公司需要创建docker-swarm集群，并测试集群的服务状态。选择Nginx服务部署到多个主机集群上进行测试docker-swarm集群。

任务描述

本任务学习创建docker-swarm集群，部署Nginx服务集群，测试Nginx服务扩容和缩容，节点失败后重新调度和集群数据卷挂载。

任务目标

◎ 能够根据环境创建docker-swarm集群
◎ 能够用命令查看docker-swarm集群工作状态
◎ 能够在docker-swarm集群中测试部署Nginx服务
◎ 能够根据需要对集群服务进行扩容、缩容、重新调度和集群数据卷挂载

任务实施

一、创建 swarm 集群

本任务创建swarm集群，通过先创建manager节点，然后将worker节点加入到集群中。

1. 环境设计

本任务采用3台宿主机，分别是：blockchain，作为服务manager主机；ubuntu-chain0和ubuntu-chain1主机作为worker主机。swarm集群架构如图7-11所示。

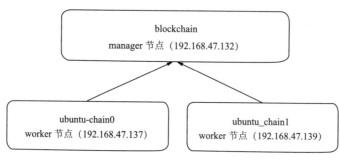

图7-11 swarm集群架构

2. 创建集群

在Blockchain主机上运行docker swarm init --advertise-addr 192.168.47.132命令，指定manager主机，生成token。

```
root@blockchain:~# docker swarm init --advertise-addr 192.168.47.132
Swarm initialized: current node (9paj3s5qmp3p8t7ogwokobmtu) is now a manager.

To add a worker to this swarm, run the following command:

    docker swarm join --token SWMTKN-1-3hs612sa7oeijw7gb9vcrwa4x480f8hjrc0da5wvttzr7wd6uw-20d7qc1v4t1viwg4r5sb0z4r4 192.168.47.132:2377

To add a manager to this swarm, run 'docker swarm join-token manager' and follow the instructions.
```

上面命令执行后，该机器自动加入到swarm集群，并创建一个集群token，获取全球唯一的 token，作为集群唯一标识。命令的输出包含了其他节点如何加入集群的命令。其中，--advertise-addr参数表示其他swarm中的worker节点使用此ip地址与manager联系。

3. 查看 swarm 集群信息

（1）查看swarm信息

使用 docker info命令可以查看docker的详细信息，其中有swarm的信息，明确看到swarm的运行状态、节点类型和数量、网络信息、manager节点的IP地址、CA证书服务器等，显示信息中进行了蓝色显示。

```
root@blockchain:~# docker info
Client:
 Debug Mode: false
Server:
 ...    （此处省略了部分显示信息）
```

```
 Plugins:
  Volume: local
  Network: bridge host ipvlan macvlan null overlay
  Log: awslogs fluentd gcplogs gelf journald json-file local logentries
splunk syslog
 Swarm: active
  NodeID: 9paj3s5qmp3p8t7ogwokobmtu
  Is Manager: true
  ClusterID: xhi8f3ze7banl5mp6k4uix8ok
  Managers: 1
  Nodes: 1
  Default Address Pool: 10.0.0.0/8
  SubnetSize: 24
  ...     (此处省略了部分显示信息)
  Root Rotation In Progress: false
  Node Address: 192.168.47.132
  Manager Addresses:
   192.168.47.132:2377
 Runtimes: runc
 ...     (此处省略了部分显示信息)
 Insecure Registries:
  192.168.47.132:5000
  127.0.0.0/8
 Registry Mirrors:
  https://y6akxxyg.mirror.aliyuncs.com/
 Live Restore Enabled: false
WARNING: No swap limit support
```

(2) 查看集群中的节点

```
root@blockchain:~# docker node ls
 ID                        HOSTNAME        STATUS      AVAILABILITY
MANAGER STATUS        ENGINE VERSION
 9paj3s5qmp3p8t7ogwokobmtu *   blockchain      Ready       Active
Leader           19.03.8
```

从命令执行的效果上看，使用docker node ls可以查看节点的ID、名称状态等信息，当前只有blockchain主机处于活动状态。下面将ubuntu-chain0和ubuntu-chain1主机加入到集群中。

4. 修改 ubuntu-chain0 和 ubuntu-chain1 主机的 docker 配置文件

在Docker的启动选项里，还原为默认选项。

(1) 修改ubuntu-chain0主机的docker配置文件并重启

先备份 /lib/systemd/system/docker.service文件。

```
root@ubuntu-chain0:~# cp /lib/systemd/system/docker.service /lib/systemd/system/docker.service.bk
```

修改/lib/systemd/system/docker.service文件，如下蓝色显示部分，修改完后保存。

[Service]

…… （此处省略了部分显示信息）

ExecStart=/usr/bin/dockerd -H fd:// --containerd=/run/containerd/containerd.sock

重新加载docker服务。

root@ubuntu-chain0:~# systemctl daemon-reload
root@ubuntu-chain0:~# systemctl restart docker

（2）修改ubuntu-chain1主机的docker配置文件并重启

用同样的方法先备份和修改。

root@ubuntu-chain1:~# cp /lib/systemd/system/docker.service /lib/systemd/system/docker.service.bk

修改/lib/systemd/system/docker.service文件，如下蓝色显示部分，修改完后保存。

[Service]

… （此处省略了部分显示信息）

ExecStart=/usr/bin/dockerd -H fd:// --containerd=/run/containerd/containerd.sock

重新加载docker服务。

root@ubuntu-chain1:~# systemctl daemon-reload
root@ubuntu-chain1:~# systemctl restart docker

5. 将worker节点加入到集群

运用docker swarm join --token token名称 192.168.47.132:2377将其他节点加入到集群中。

（1）将ubuntu-chain0主机（192.168.47.137）加入到集群中

root@ubuntu-chain0:~# docker swarm join --token SWMTKN-1-3hs612sa7oeijw7gb9vcrwa4x480f8hjrc0da5wvttzr7wd6uw-20d7qc1v4t1viwg4r5sb0z4r4 192.168.47.132:2377
This node joined a swarm as a worker.

从命令显示信息看ubuntu-chain0主机已作为worker加入到集群中了。

（2）将ubuntu-chain1主机（192.168.47.139）加入到集群中

root@ubuntu-chain1:~# docker swarm join --token SWMTKN-1-3hs612sa7oeijw7gb9vcrwa4x480f8hjrc0da5wvttzr7wd6uw-20d7qc1v4t1viwg4r5sb0z4r4 192.168.47.132:2377
This node joined a swarm as a worker.

从命令显示信息看ubuntu-chain1主机也已作为worker加入到集群中了。

6. 查看集群节点

再次用docker node ls命令查看集群节点情况。可以看到STATUS状态均为Ready，AVAILABILITY均为Active，其中blockchain节点的MANAGER STATUS状态是Leader，说明是manager角色。

root@blockchain:~# docker node ls
ID   HOSTNAME   STATUS   AVAILABILITY

```
MANAGER STATUS        ENGINE VERSION
  9paj3s5qmp3p8t7ogwokobmtu *   blockchain          Ready          Active
Leader              19.03.8
  ysea0nodbom4qg3jvs5g66po9     ubuntu-chain0       Ready          Active
19.03.8
  rq166pl89u2ncoerhr0y3ltsr     ubuntu-chain1       Ready          Active
19.03.8
```

二、部署 nginx 集群服务

下面任务中将创建3个容器实例，swarm 会自动将容器进行分配给集群节点。

1. 创建 overlay 网络

创建一个 my-nginxnet0 网络。

```
root@blockchain:~# docker network create --driver overlay my-nginxnet0
asf04xaxht3u4tsq2jmgstwjm
```

创建完成后进行查看，可以看到创建了驱动是 overlay 的 my-nginxnet0 的网络。

```
root@blockchain:~# docker network ls
NETWORK ID          NAME                DRIVER              SCOPE
…（此处省略了部分显示信息）
asf04xaxht3u        my-nginxnet0        overlay             swarm
…（此处省略了部分显示信息）
```

2. 创建 nginx 集群

在 blockchain 主机上创建 nginx 集群，用 docker service create 命令实现，服务的名字为 my_nginx。

```
root@blockchain:~# docker service create --replicas 3 --network my-nginxnet0 --name my_nginx -p 80:80 nginx
i7pz9otg3aj8i4fmq3y1j0cr9
overall progress: 3 out of 3 tasks
1/3: running   [==================================================>]
2/3: running   [==================================================>]
3/3: running   [==================================================>]
verify: Service converged
```

该命令用 --replicas 3 指明副本数量，即运行几个 nginx 容器副本，用 --network my-nginxnet0 指明使用的网络，--name my_nginx 指明容器的名称，-p 80:80 指明端口映射。

3. 查看 nginx 集群信息

（1）查看 nginx 容器实例在主机上运行状况

根据命令执行结果显示，可以看到3个实例（my_nginx.1、my_nginx.2、my_nginx.3）分别运行在3个节点上。

```
root@blockchain:~# docker service ps my_nginx
ID                  NAME                IMAGE               NODE
DESIRED STATE       CURRENT STATE       ERROR               PORTS
  mj5692rsl8lv        my_nginx.1          nginx:latest        blockchain
```

```
Running                Running 32 minutes ago
  qnqk29r2fwes         my_nginx.2            nginx:latest          ubuntu-chain1
Running                Running 28 minutes ago
  8xh47ims09w6         my_nginx.3            nginx:latest          ubuntu-chain0
Running                Running 32 minutes ago
```

(2) 在每个节点上查看nginx容器实例运行状况

分别在blockchain、ubuntu-chain0、ubuntu-chain1主机上查看nginx容器实例。

① 在blockchain主机上查看。

```
root@blockchain:~# docker ps
  CONTAINER ID         IMAGE                 COMMAND               CREATED
STATUS                 PORTS                 NAMES
  c67907a6c867         nginx:latest          "nginx -g 'daemon of…"   32 minutes ago
Up 12 minutes          80/tcp                my_nginx.1.mj5692rsl8lv7xli1vbarzjkv
```

② 在ubuntu-chain0主机上查看。

```
root@ubuntu-chain0:~# docker ps
  CONTAINER ID         IMAGE                 COMMAND               CREATED
STATUS                 PORTS                 NAMES
  8e3d8c2bc0a3         nginx:latest          "nginx -g 'daemon of…"   32 minutes ago
Up 12 minutes          80/tcp                my_nginx.3.8xh47ims09w6yvb73ys4o14wx
```

③ 在ubuntu-chain1主机上查看。

```
root@ubuntu-chain1:~# docker ps
  CONTAINER ID         IMAGE                 COMMAND               CREATED
STATUS                 PORTS                 NAMES
  78057746a368         nginx:latest          "nginx -g 'daemon of…"   28 minutes ago
Up 8 minutes           80/tcp                my_nginx.2.qnqk29r2fwes0oftu6uqeztfy
```

从上面查看结果可以看出，运行的3个副本平均被分配到3台主机上。

(3) 在blockchain主机上查看swarm服务信息

在manager节点上用docker service ls命令查看swarm服务信息，可以看到ID、名称、模式、副本情况、镜像和端口映射信息，其中可以看到总数量是3，有3个副本正在运行，与上一条命令查看的结果一致。

```
root@blockchain:~# docker service ls
ID              NAME         MODE          REPLICAS    IMAGE           PORTS
i7pz9otg3aj8    my_nginx     replicated    3/3         nginx:latest    *:80->80/tcp
```

(4) 查看my_nginx服务的详细信息

使用docker service inspect命令显示服务的详细信息，主要有服务的ID、该服务创建（CreatedAt）和更新（UpdatedAt）的时间、资源使用策略（Resources）、重启策略（RestartPolicy）、节点操作系统平台（Placement）、副本数量（Replicated）、更新配置（UpdateConfig）、回滚配置（RollbackConfig）、服务IP和端口配置信息。

```
root@blockchain:~# docker service inspect my_nginx
[
```

```
{
    "ID": "i7pz9otg3aj8i4fmq3y1j0cr9",
    "Version": {
        "Index": 34
    },
    "CreatedAt": "2020-04-27T12:59:32.174843107Z",
    "UpdatedAt": "2020-04-27T12:59:32.177617674Z",
    "Spec": {
        "Name": "my_nginx",
        "Labels": {},
        "TaskTemplate": {
            "ContainerSpec": {
                "Image": "nginx:latest@sha256:86ae264c3f4acb99b2dee4d0098c40cb8c46dcf9e1148f05d3a51c4df6758c12",
                "Init": false,
                "StopGracePeriod": 10000000000,
                "DNSConfig": {},
                "Isolation": "default"
            },
            "Resources": {
                "Limits": {},
                "Reservations": {}
            },
            "RestartPolicy": {
                "Condition": "any",
                "Delay": 5000000000,
                "MaxAttempts": 0
            },
            "Placement": {
                "Platforms": [
                    {
                        "Architecture": "amd64",
                        "OS": "linux"
                    },
                    {
                        "OS": "linux"
                    },
                    {
                        "Architecture": "arm64",
                        "OS": "linux"
                    },
                    {
                        "Architecture": "386",
```

```json
                    "OS": "linux"
                },
                {
                    "Architecture": "ppc64le",
                    "OS": "linux"
                },
                {
                    "Architecture": "s390x",
                    "OS": "linux"
                }
            ]
        },
        "Networks": [
            {
                "Target": "asf04xaxht3u4tsq2jmgstwjm"
            }
        ],
        "ForceUpdate": 0,
        "Runtime": "container"
    },
    "Mode": {
        "Replicated": {
            "Replicas": 3
        }
    },
    "UpdateConfig": {
        "Parallelism": 1,
        "FailureAction": "pause",
        "Monitor": 5000000000,
        "MaxFailureRatio": 0,
        "Order": "stop-first"
    },
    "RollbackConfig": {
        "Parallelism": 1,
        "FailureAction": "pause",
        "Monitor": 5000000000,
        "MaxFailureRatio": 0,
        "Order": "stop-first"
    },
    "EndpointSpec": {
        "Mode": "vip",
        "Ports": [
            {
```

```
                            "Protocol": "tcp",
                            "TargetPort": 80,
                            "PublishedPort": 80,
                            "PublishMode": "ingress"
                        }
                    ]
                }
            },
            "Endpoint": {
                "Spec": {
                    "Mode": "vip",
                    "Ports": [
                        {
                            "Protocol": "tcp",
                            "TargetPort": 80,
                            "PublishedPort": 80,
                            "PublishMode": "ingress"
                        }
                    ]
                },
                "Ports": [
                    {
                        "Protocol": "tcp",
                        "TargetPort": 80,
                        "PublishedPort": 80,
                        "PublishMode": "ingress"
                    }
                ],
                "VirtualIPs": [
                    {
                        "NetworkID": "rtimbrsph4pfnwe7800ol0hik",
                        "Addr": "10.0.0.5/24"
                    },
                    {
                        "NetworkID": "asf04xaxht3u4tsq2jmgstwjm",
                        "Addr": "10.0.1.2/24"
                    }
                ]
            }
        }
    ]
```

4. 测试集群扩缩容

(1) 浏览器访问测试

在浏览器中输入manager管理节点的IP地址，可以测试网页信息，如图7-12所示。

图 7-12 manager 管理节点网页信息

输入其他worker节点的IP地址也可以访问该网页。ubuntu-chain0节点的访问情况如图7-13所示。

图 7-13 ubuntu-chain0 节点网页信息

在ubuntu-chain1节点上也可以访问该网页，如图7-14所示。

图 7-14 ubuntu-chain1 节点网页信息

从以上测试结果可以看出，无论访问集群中的哪个节点，网页的内容都是一样的，起到了集群相互冗余作用。

(2) 将manager节点设置不指派任务

一般情况下manager节点用来管理工作，不承担具体业务，可以将manager节点设置为drain状态，这样当任务指派时就不会将任务分配给manager节点了。

```
root@blockchain:~# docker node update --availability drain blockchain
blockchain
```

执行完命令后，查看各节点的状态，发现blockchain的状态已经改为Drain，而不是以前的Active了。

```
root@blockchain:~# docker node ls
ID                           HOSTNAME           STATUS    AVAILABILITY   MANAGER STATUS   ENGINE VERSION
9paj3s5qmp3p8t7ogwokobmtu *  blockchain         Ready     Drain          Leader           19.03.8
ysea0nodbom4qg3jvs5g66po9    ubuntu-chain0      Ready     Active                          19.03.8
rq166pl89u2ncoerhr0y3ltsr    ubuntu-chain1      Ready     Active                          19.03.8
```

进一步查看服务的节点状态，发现3个副本所在的宿主机也发生变化了，这时，在ubuntu-chain0节点上面运行一个实例副本，在ubuntu-chain1节点上面运行两个实例副本，也就是manager会根据调度算法，在ubuntu-chain0和ubuntu-chain1节点上进行调度，保证有3个实例副本运行。

```
root@blockchain:~# docker service ps my_nginx -f DESIRED-STATE=Running
ID             NAME          IMAGE          NODE            DESIRED STATE   CURRENT STATE            ERROR   PORTS
aj1z5zzn75ib   my_nginx.1    nginx:latest   ubuntu-chain1   Running         Running 7 minutes ago
l6ru8d8sbwmb   my_nginx.2    nginx:latest   ubuntu-chain1   Running         Running 7 minutes ago
6nzi3f0xx50p   my_nginx.3    nginx:latest   ubuntu-chain0   Running         Running 41 minutes ago
```

（3）测试缩容

有时候因为业务量的变化，为企业减少成本，当业务流量不高时，可以适当缩容，本任务将nginx集群中实例副本的数量减缩为两个。可以采用docker service scale命令或者docker service update --replicas均可实现。下面用docker service scale命令进行测试。

```
root@blockchain:~# docker service scale my_nginx=2
my_nginx scaled to 2
overall progress: 2 out of 2 tasks
1/2: running   [==================================================>]
2/2: running   [==================================================>]
verify: Service converged
```

从命令执行过程可以看出，已将my_nginx服务缩减为两个实例副本了。用docker service ps查看，可以看出两个实例副本分别运行在ubuntu-chain0和ubuntu-chain1节点上。

```
root@blockchain:~# docker service ps my_nginx -f DESIRED-STATE=Running
    ID              NAME            IMAGE           NODE
DESIRED STATE       CURRENT STATE         ERROR           PORTS
    aj1z5zzn75ib    my_nginx.1      nginx:latest    ubuntu-chain1    Running
Running 17 minutes ago
    6nzi3f0xx50p    my_nginx.3      nginx:latest    ubuntu-chain0    Running
Running 52 minutes ago
```

(4) 测试扩容

将nginx集群扩容为5个副本，这次采用docker service update --replicas命令测试，用docker service scale命令也同样能实现。

```
root@blockchain:~# docker service update --replicas=5 my_nginx
my_nginx
overall progress: 5 out of 5 tasks
1/5: running   [==================================================>]
2/5: running   [==================================================>]
3/5: running   [==================================================>]
4/5: running   [==================================================>]
5/5: running   [==================================================>]
verify: Service converged
```

从命令执行过程可以看出，已经将my_nginx服务扩容为5个实例副本了。用docker service ps查看，可以看出5个实例副本分别运行在ubuntu-chain0和ubuntu-chain1节点上，其中ubuntu-chain0节点运行了3个实例副本，ubuntu-chain1节点运行了两个实例副本。

```
root@blockchain:~# docker service ps my_nginx -f DESIRED-STATE=Running
    ID              NAME            IMAGE           NODE
DESIRED STATE       CURRENT STATE         ERROR           PORTS
    aj1z5zzn75ib    my_nginx.1      nginx:latest    ubuntu-chain1    Running
Running 23 minutes ago
    6nmi3jehnt17    my_nginx.2      nginx:latest    ubuntu-chain1    Running
Running 15 seconds ago
    6nzi3f0xx50p    my_nginx.3      nginx:latest    ubuntu-chain0    Running
Running 58 minutes ago
    wwymiiw39f37    my_nginx.4      nginx:latest    ubuntu-chain0    Running
Running 14 seconds ago
    i8y1z4sdzzlt    my_nginx.5      nginx:latest    ubuntu-chain0    Running
Running 13 seconds ago
```

5. 测试 nginx 集群的容灾性

集群的优势就是提高性能并增强可用性，当集群中某些节点发生意外死机时，manager会根据调度算法马上在其他节点上增加实例副本。

(1) 测试死机状况

本任务测试将ubuntu-chain0节点死机，将ubuntu-chain0节点的docker服务关闭，然后查看容器的调度情况。

① 关闭ubuntu-chain0节点的docker服务。

```
root@ubuntu-chain0:~# systemctl stop docker
```

查看docker运行状况，可以看到docker服务已经关闭，处于不活动状态。

```
root@ubuntu-chain0:~# systemctl status docker
  docker.service - Docker Application Container Engine
   Loaded: loaded (/lib/systemd/system/docker.service; enabled; vendor preset: enabled)
   Active: inactive (dead) since Tue 2020-04-28 12:55:47 UTC; 10s ago
     Docs: https://docs.docker.com
  Process: 1144 ExecStart=/usr/bin/dockerd -H fd:// --containerd=/run/containerd/containerd.sock (code=exited, status=0/SUCCESS)
 Main PID: 1144 (code=exited, status=0/SUCCESS)
 ...（此处省略了部分显示信息）
```

② 查看my_nginx服务运行状况。

关闭ubuntu-chain0节点的docker服务后，用命令docker service ps my_nginx查看服务运行状况。从命令执行结果可以看出，已经将my_nginx服务的5个实例副本调度到ubuntu-chain1节点上了。

```
root@blockchain:~# docker service ps my_nginx -f DESIRED-STATE=Running
ID              NAME          IMAGE          NODE            DESIRED STATE   CURRENT STATE            ERROR    PORTS
aj1z5zzn75ib    my_nginx.1    nginx:latest   ubuntu-chain1   Running         Running 38 minutes ago
6nmi3jehntl7    my_nginx.2    nginx:latest   ubuntu-chain1   Running         Running 14 minutes ago
0rj45fazecbo    my_nginx.3    nginx:latest   ubuntu-chain1   Running         Running 3 minutes ago
2o3u6fyse131    my_nginx.4    nginx:latest   ubuntu-chain1   Running         Running 3 minutes ago
y4z397swa639    my_nginx.5    nginx:latest   ubuntu-chain1   Running         Running 3 minutes ago
root@blockchain:~#
```

③ 在ubuntu-chain1节点上查看容器运行状况。

同样可以在ubuntu-chain1节点上查看5个实例的运行情况。

```
root@ubuntu-chain1:~# docker ps
CONTAINER ID    IMAGE          COMMAND                 CREATED         STATUS          PORTS     NAMES
cfe3cecf1e2c    nginx:latest   "nginx -g 'daemon of…"  7 minutes ago   Up 7 minutes    80/tcp    my_nginx.5.y4z397swa6399k0l03jyk2db7
2e639189800d    nginx:latest   "nginx -g 'daemon of…"  7 minutes ago   Up 7 minutes    80/tcp    my_nginx.3.0rj45fazecbo428ob4te2ry91
```

```
    b92d8d6252a4        nginx:latest        "nginx -g 'daemon of…"    7 minutes ago
Up 7 minutes            80/tcp              my_nginx.4.2o3u6fyse131cei88h1pcgxof
    cc6f8395fe30        nginx:latest        "nginx -g 'daemon of…"    18 minutes ago
Up 18 minutes           80/tcp              my_nginx.2.6nmi3jehntl7151akaeel9bbk
    4cc1483858d9        nginx:latest        "nginx -g'daemon of…"    41 minutes ago
Up 41 minutes           80/tcp              my_nginx.1.aj1z5zzn75ib0bsfjdo12y7r9
```

(2) 测试节点恢复状况

上面通过关闭ubuntu-chain0节点的docker服务模拟ubuntu-chain0节点的死机，下面将ubuntu-chain0节点的docker服务启动模拟ubuntu-chain0节点的恢复，测试容器的调度情况。

① 启动ubuntu-chain0节点的docker服务。

```
root@ubuntu-chain0:~# systemctl start docker
```

启动docker服务后，可以看到docker服务已经开启，处于活动状态。

```
root@ubuntu-chain0:~# systemctl status docker
  docker.service - Docker Application Container Engine
    Loaded: loaded (/lib/systemd/system/docker.service; enabled; vendor preset: enabled)
    Active: active (running) since Tue 2020-04-28 13:14:29 UTC; 8s ago
      Docs: https://docs.docker.com
  Main PID: 5306 (dockerd)
     Tasks: 16
    CGroup: /system.slice/docker.service
            └─5306 /usr/bin/dockerd -H fd:// --containerd=/run/containerd/containerd.sock
… （此处省略了部分显示信息）
```

② 查看my_nginx服务运行状况。

从命令执行结果可以看出，my_nginx服务的5个实例副本仍然在ubuntu-chain1节点上运行，并没有重新调度到ubuntu-chain1，也就是在默认情况下，swarm并不会将容器调度到曾失效的节点。

```
root@blockchain:~# docker service ps my_nginx -f DESIRED-STATE=Running
ID                  NAME                IMAGE               NODE
DESIRED STATE       CURRENT STATE            ERROR              PORTS
    aj1z5zzn75ib        my_nginx.1          nginx:latest        ubuntu-chain1
Running             Running 55 minutes ago
    6nmi3jehntl7        my_nginx.2          nginx:latest        ubuntu-chain1
Running             Running 31 minutes ago
    0rj45fazecbo        my_nginx.3          nginx:latest        ubuntu-chain1
Running             Running 20 minutes ago
    2o3u6fyse131        my_nginx.4          nginx:latest        ubuntu-chain1
Running             Running 20 minutes ago
    y4z397swa639        my_nginx.5          nginx:latest        ubuntu-chain1
Running             Running 20 minutes ago
```

③ 重新更新使容器重新调度。

下面测试在更新命令中使用--env-add reschedule:on-node-failure选项进行更新，使集群重新调度。

```
root@blockchain:~# docker service update --env-add reschedule:on-node-
failure my_nginx
my_nginx
overall progress: 5 out of 5 tasks
1/5: running   [==================================================>]
2/5: running   [==================================================>]
3/5: running   [==================================================>]
4/5: running   [==================================================>]
5/5: running   [==================================================>]
verify: Service converged
```

④ 重新查看my_nginx服务运行状况。

从命令执行结果可以看出，my_nginx服务的5个实例副本重新调度，其中ubuntu-chain0节点上运行两个实例副本，ubuntu-chain1节点运行3个实例副本，可见运用--env-add reschedule:on-node-failure选项进行更新，可以实现集群重新调度。

```
root@blockchain:~# docker service ps my_nginx -f DESIRED-STATE=Running
ID              NAME            IMAGE           NODE            
DESIRED STATE       CURRENT STATE       ERROR           PORTS
  ywhvczcu34tj      my_nginx.1      nginx:latest    ubuntu-chain1   Running
Running about a minute ago
  oxjw4m2reqos      my_nginx.2      nginx:latest    ubuntu-chain0   Running
Running 2 minutes ago
  ye5fog5f176m      my_nginx.3      nginx:latest    ubuntu-chain1   Running
Running about a minute ago
  cile3x139vne      my_nginx.4      nginx:latest    ubuntu-chain1   Running
Running 2 minutes ago
  yi6lxriuwior      my_nginx.5      nginx:latest    ubuntu-chain0   Running
Running 2 minutes ago
```

6. 测试 nginx 集群数据卷挂载

在ubuntu-chain0和ubuntu-chain1节点上分别创建网页文件夹存放网页，然后挂载到nginx容器中测试nginx数据卷的挂载。

（1）在ubuntu-chain0节点上创建目录并创建网页

```
root@ubuntu-chain0:~# mkdir testswarm
root@ubuntu-chain0:~# cd testswarm/
root@ubuntu-chain0:~/testswarm# mkdir nginxhtml
root@ubuntu-chain0:~/testswarm# cd nginxhtml/
root@ubuntu-chain0:~/testswarm/nginxhtml# echo 'this is a swarm test page' >index.html
```

（2）在ubuntu-chain1节点上创建目录并创建网页

```
root@ubuntu-chain1:~# mkdir testswarm
root@ubuntu-chain1:~# cd testswarm/
root@ubuntu-chain1:~/testswarm# mkdir nginxhtml
root@ubuntu-chain1:~/testswarm# cd nginxhtml
root@ubuntu-chain1:~/testswarm/nginxhtml# echo 'this is a swarm test page' >index.html
```

（3）在blockchain节点上进行更新

使用docker service update命令进行更新，用--mount-add选项增加卷，用type=bind设置类型，用source=/root/testswarm/nginxhtml设置源，用target=/usr/share/nginx/html设置容器的目标目录。

```
root@blockchain:~# docker service update  --mount-add type=bind,source=/root/testswarm/nginxhtml,target=/usr/share/nginx/html my_nginx
my_nginx
overall progress: 5 out of 5 tasks
1/5: running   [==================================================>]
2/5: running   [==================================================>]
3/5: running   [==================================================>]
4/5: running   [==================================================>]
5/5: running   [==================================================>]
verify: Service converged
```

（4）查看更新后的nginx集群状态

从命令执行结果上看，my_nginx服务的5个实例副本在运行，其中ubuntu-chain0节点上运行2个实例副本，ubuntu-chain1节点运行3个实例副本。

```
root@blockchain:~# docker service ps my_nginx -f DESIRED-STATE=Running
ID              NAME              IMAGE             NODE
DESIRED STATE        CURRENT STATE         ERROR           PORTS
  yjjg2j6mku64      my_nginx.1        nginx:latest      ubuntu-chain1
Running              Running 3 minutes ago
  mh4d6xlprs3m      my_nginx.2        nginx:latest      ubuntu-chain0
Running              Running 3 minutes ago
  mq5ihum8isyl      my_nginx.3        nginx:latest      ubuntu-chain0
Running              Running 3 minutes ago
  kmim5smp7i27      my_nginx.4        nginx:latest      ubuntu-chain1
Running              Running 3 minutes ago
  ur4vn27wlh60      my_nginx.5        nginx:latest      ubuntu-chain1
Running              Running 3 minutes ago
```

（5）测试网页

在浏览器上输入集群中的节点的IP地址，输出结果如下：

① 查看192.168.47.132节点，如图7-15所示。

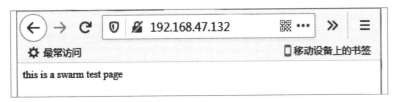

图 7-15　192.168.47.132 节点网页

② 查看192.168.47.137节点，如图7-16所示。

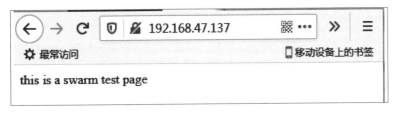

图 7-16　192.168.47.137 节点网页

③ 查看192.168.47.139节点，如图7-17所示。

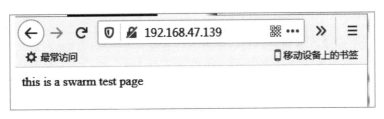

图 7-17　192.168.47.139 节点网页

从命令执行结果可以看出，无论访问哪个节点，均能查看到正常的网页显示，数据卷挂载成功。

7. 删除 my_nginx 服务

```
root@blockchain:~/stack-lnmp# docker service rm my_nginx
my_nginx
```

至此，完成nginx的集群服务，实现高可用的Web服务，完成后删除my_nginx服务，释放节点资源。

任务总结

① 要理解swarm的架构，清楚manger和worker的角色功能，manger起到注册服务及服务发现功能，worker主要承担业务计算服务功能。

② 要理解掌握swarm中task、service、stack的概念。task是对应worker节点上的一个容器，提供的是单一服务，如在某个worker节点上运行nginx容器，提供Web服务；services可能包含多个task，以集群的方式提供某一个单一服务，如nginx服务在多个worker节点上运行，形成一个集群，任何一个worker节点死机不影响集群提供Web服务；stack是服务栈的概念，是包含多个提供不同功能的service组成，运用stack提供了综合性的服务功能，如有多个service组成综合动态网

站，分别由nginx、php、mysql几个service组成。

③ manager具有注册和服务发现功能，可以将多个worker节点统一管理，用一个虚拟的IP代替每个worker节点提供服务，任何worker节点故障，对外是透明的。例如，本任务中虚拟IP地址是10.0.0.5/24和10.0.1.2/24，代表容器eth0和eth1的地址。

④ 扩缩容可以采用docker service scale命令或者docker service update --replicas实现，利用docker service update 命令更新时使用--env-add reschedule:on-node-failure 选项可以发生重新调度。

任务扩展

本任务中通过docker-swarm部署nginx服务，读者可以自行测试部署Apache进行测试。

任务三 运用 Docker Stack 部署 LNMP 网站

任务场景

随着公司的业务需求，要保证公司的业务的可靠性，并能提供网站网页文件的持久性和容易更改，需要将前期的LNMP网站部署到多主机环境中，技术部门确定采用docker-swarm部署LNMP网站集群，并测试集群的运行状态。

任务描述

本任务学习采用docker-swarm部署LNMP网站集群，为保证网站网页文件的持久性，将网站的网页文件通过NFS服务器挂载到容器中。

任务目标

◎ 能根据需求设计物理网络架构
◎ 能根据需求编写stack file文件
◎ 能根据需求设置nfs服务共享
◎ 能部署LNMP网站并用命令查看docker-swarm集群工作状态
◎ 能测试LNMP网站集群工作状态

任务实施

单机模式下，可以使用Docker Compose来编排多个服务，但Docker Compose只能实现在单个主机上简单部署。在多主机环境中，Docker Stack可以实现多服务的部署，需要使用stack file配置文件。stack file是一种yaml格式的文件，类似于 docker-compose.yml 文件，它定义了一个或多个服务，并定义了服务的环境变量、部署标签、容器数量及相关的环境特定配置等。但Docker stack与compose相比其不支持build、links和network_mod等（随着版本升级也在发生变化），Docker stack file有一个新的指令deploy。

主要区别如下：

① stack命令不支持build。

② docker compose不支持 deploy。

③ docker stack是使用 Go 语言创建的内建命令。

④ docker compose是使用Python语言构建的第三方命令。

⑤ docker stack是swarm mode的一部分，即使是单机使用，也需要一个swarm节点。

⑥ docker stack只能支持version3 以上版本。

一、环境设计

1. 架构设计

采用图7-18所示架构，LNMP由3个容器组成，分别是nginx、php和mysql。3个容器的命名也分别用nginx、php和mysql，3个容器分别运行在ubuntu-chain0和ubuntu-chain1节点上，并每个服务运行两个实例副本。

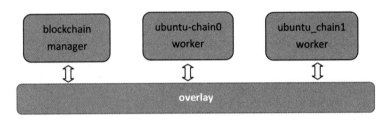

图 7-18　部署 LNMP 网站架构

2. 目录结构设计

创建stack-lnmp目录，并逐步创建如下目录结构和编辑相关文档：

```
root@ubuntu-chain0:~/stack-lnmp# tree
.
├── mysql
│   ├── conf
│   │   └── my.cnf
│   └── data
├── nginx
│   └── nginx.conf
└── wwwroot
    └── index.php
```

同样在ubuntu-chain1节点上也创建相应文件和目录。

二、编辑配置文件

1. 编辑 nginx.conf 配置文件

在nginx文件夹中编辑nginx.conf文件，重点是修改location中的内容，修改数据定义nginx和PHP连接的配置，该文件完全可以参考compose部分的LNMP网站建设内容。分别在ubuntu-chain0和ubuntu-chain1上编辑该文件。

```
root@ubuntu-chain0:~/stack-lnmp/nginx# vi nginx.conf
...　　（此处省略了部分显示信息）
```

```
        server {
            listen 80;
            server_name localhost;
            #指定网页所在的目录
    root /usr/share/nginx/html;
            index index.html index.php;
            location ~ .*\.php$ {
            #指明php套接字,其中php-cgi是docker-compose.yml文件中定义的PHP别名。
            fastcgi_pass php:9000;
                fastcgi_index index.php;        #指明网页文件
                #以下两行指明PHP脚本文件名所在的路径和文件名,其中include是参数定义。
                fastcgi_param SCRIPT_FILENAME /usr/share/nginx/html/$fastcgi_script_name;
                include fastcgi_params;
            }
        }
        #gzip  on;
        include /etc/nginx/conf.d/*.conf;
    }
```

2. 编辑 mysql 配置文件

分别在ubuntu-chain0和ubuntu-chain1上编辑该文件。

```
root@ubuntu-chain0:~/stack-lnmp/mysql/conf# vi my.cnf
[mysqld]
user=mysql
port=3306
datadir=/var/lib/mysql
socket=/var/run/mysqld/mysqld.sock
pid-file=/var/run/mysqld/mysqld.pid
#log_error=/var/log/mysql/error.log
character_set_server = utf8
max_connections=3600
```

3. 编辑网页 index.php 文件

分别在ubuntu-chain0和ubuntu-chain1节点上wwwroot文件夹中编辑index.php网页文件,本次测试打印PHP基本信息。

```
root@ubuntu-chain0:~/stack-lnmp# echo '<?php phpinfo(); ?>' > index.php
```

4. 编辑 docker-compose.yml 文件

按照如下内容编辑docker-compose.yml文件

```
# 指定服务版本号
version: '3.8'
# 服务
services:
```

```yaml
# 服务名称
  mysql:
  # 指定服务容器名字
      hostname: mysql
  # 指定使用官方mysql5.6版本
      image: mysql
  # 映射端口
      ports:
        - 3306:3306
  # 映射服务数据卷路径
      volumes:
        - ./mysql/conf:/etc/mysql/conf.d
        - ./mysql/data:/var/lib/mysql
  # 指定数据库变量
      environment:
      # 设置数据库密码，这里用123456，用户根据需要更改
        MYSQL_ROOT_PASSWORD: 123456
      # 添加user用户
        MYSQL_USER: user
      # 设置user用户密码
        MYSQL_PASSWORD: user123
      networks:
        - lnmp
      deploy:
        replicas: 2
        update_config:
          parallelism: 2
          delay: 10s
        restart_policy:
          condition: on-failure
# 服务名称
  php:
  # 指定服务容器名字
      hostname: php
  # 构建
      image: bitnami/php-fpm
      # 映射mysql服务别名
      ports:
           - 9000:9000
  # 映射服务数据卷路径
      volumes:
           - ./wwwroot:/usr/share/nginx/html
      networks:
```

```yaml
              - lnmp
        stdin_open: true
        tty: true
        deploy:
          replicas: 2
          update_config:
            parallelism: 2
            delay: 10s
          restart_policy:
            condition: on-failure
  # 服务名称
     nginx:
     # 指定服务容器名字
        hostname: nginx
        # 构建
        image: nginx
        # 映射数组级的端口
        ports:
                - 80:80
                - 443:443
        # 映射服务数据卷路径
        volumes:
                - ./wwwroot:/usr/share/nginx/html
                - ./nginx/nginx.conf:/etc/nginx/nginx.conf
        networks:
                - lnmp
        deploy:
          replicas: 2
          update_config:
            parallelism: 2
            delay: 10s
          restart_policy:
            condition: on-failure
  networks:
      lnmp:
        driver: overlay
```

三、利用 docker stack deplay 部署 LNMP 和测试

1. 执行 docker stack deploy 命令一键部署 LNMP

```
root@blockchain:~/stack-lnmp# docker stack deploy -c docker-compose.yml stack-lnmp
Creating network stack-lnmp_lnmp
Creating service stack-lnmp_mysql
```

```
Creating service stack-lnmp_nginx
Creating service stack-lnmp_php
```

2. 查看运行状况

用docker stack ps stack-lnmp命令查看多个服务运行的节点和状态。可以看到实例副本分别运行在ubuntu-chain0和ubuntu-chain1节点上。

```
root@blockchain:~/stack-lnmp# docker stack ps stack-lnmp
ID                 NAME                  IMAGE                    NODE
DESIRED STATE      CURRENT STATE                 ERROR            PORTS
  r3h10mlre8j6     stack-lnmp_nginx.1    nginx:latest             ubuntu-chain0
Running            Running about a minute ago
  j7km3fvl29d2     stack-lnmp_php.1      bitnami/php-fpm:latest
ubuntu-chain0      Running               Running about a minute ago
  2p54mn39f61a     stack-lnmp_mysql.1    mysql:latest             ubuntu-chain0
Running            Running about a minute ago
  xnnogy4jyzqj     stack-lnmp_nginx.2    nginx:latest             ubuntu-chain1
Running            Running about a minute ago
  m2fnfh9dffoi     stack-lnmp_php.2      bitnami/php-fpm:latest
ubuntu-chain1      Running               Running about a minute ago
  okwb05rql67o     stack-lnmp_mysql.2    mysql:latest             ubuntu-chain1
Running            Preparing about a minute ago
```

用docker stack services stack-lnmp命令查看stack-lnmp副本的运行情况。

```
root@blockchain:~# docker stack services stack-lnmp
ID              NAME              MODE         REPLICAS    IMAGE            PORTS
  cw18vloodptu    stack-lnmp_mysql  replicated   2/2         mysql:latest
*:3306->3306/tcp
  ovz1x87mspnd    stack-lnmp_nginx  replicated   2/2         nginx:latest
*:80->80/tcp, *:443->443/tcp
  yabqj2glrf26    stack-lnmp_php    replicated   2/2         bitnami/php-
fpm:latest    *:9000->9000/tcp
```

3. 进行网页测试

在浏览器中输入节点主机IP地址测试，成功！

① 测试manager节点地址。图7-19所示为blockchain主机节点上的网页显示结果。

② 测试ubuntu-chain0节点。图7-20所示为ubuntu-chain0主机节点上的网页显示结果（仅截取网页上面部分）。

③ 测试ubuntu-chain1节点。图7-21所示为ubuntu-chain1主机节点上的网页显示结果（仅截取网页上面部分）。

至此，通过docker stack deplay方式部署LNMP已经完成，几个节点网页显示内容相同，起到相互冗余作用。上面的部署方式也存在缺点，该方式中网页文件分别存放在ubuntu-chain0和ubuntu-chain1节点上，并不方便维护。下面进行调整，将网页文件存放在nfs服务器上。

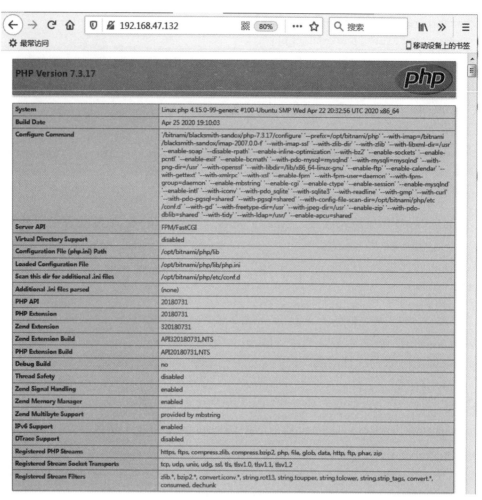

图 7–19 blockchain 节点网页显示结果

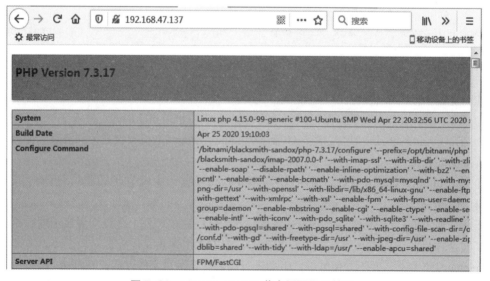

图 7–20 ubuntu–chain0 节点网页显示结果

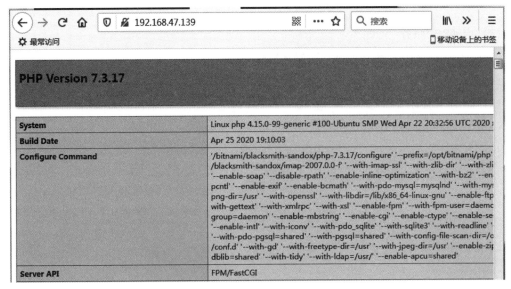

图 7-21　ubuntu-chain1 节点网页显示结果

四、本地卷映射到 nfs 服务器上实现卷的挂载

可以选择将网页文件存放在nfs服务器上，集群中的nginx服务使用nfs共享的网页内容，这样方便管理，也利于集群使用。

1. 创建 nfs 共享目录

创建共享目录并设置权限。

```
root@blockchain:~# mkdir -p nfs-share/wwwroot
root@blockchain:~# chmod 777 nfs-share/wwwroot
```

2. 安装配置 nfs 服务

本任务将nfs服务安装在blockchain主机manager节点上。

（1）安装nfs服务

```
root@blockchain:~# apt install nfs-kernel-server
```

（2）修改nfs配置文件

修改/etc/exports文件，在/etc/exports文件中添加一行参数。

```
/root/nfs-share/wwwroot 192.168.47.*(rw,sync,no_root_squash,no_subtree_check)
```

参数说明：

① /root/nfs-share/wwwroot是与nfs服务客户端共享的目录；

② 192.168.47.*代表允许192.168.47.0/24网段访问（也可以使用具体的IP）；

③ rw：挂接此目录的客户端对该共享目录具有读写权限；

④ sync：资料同步写入内存和硬盘；

⑤ no_root_squash：客户机用root访问该共享文件夹时，不映射root用户。（root_squash：客户机用root用户访问该共享文件夹时，将root用户映射成匿名用户）；

⑥ no_subtree_check：不检查父目录的权限。

3. 重新启动 nfs 服务

（1）启动RPC服务

nfs是一个RPC程序，使用它前，需要映射好端口，通过rpcbind设定。

```
root@blockchain:~# systemctl restart rpcbind
```

（2）启动nfs服务

```
root@blockchain:~# systemctl restart nfs-kernel-server
```

（3）查看注册的端口

从命令显示结果上看nfs服务端口正常开启。

```
root@blockchain:~# rpcinfo -p localhost
   program vers proto   port  service
    100000    4   tcp    111  portmapper
    100000    3   tcp    111  portmapper
    100000    2   tcp    111  portmapper
    100000    4   udp    111  portmapper
    100000    3   udp    111  portmapper
    100000    2   udp    111  portmapper
    ...    （此处省略了部分显示信息）
```

（4）设置nfs服务启动自动加载

```
root@blockchain:~/stack-lnmp# systemctl enable rpcbind nfs-kernel-server
```

（5）查看加载的目录

```
root@blockchain:~# showmount --exports
Export list for blockchain:
/root/nfs-share/wwwroot 192.168.47.*
```

从命令执行结果可以看出，/root/nfs-share/wwwroot目录成功加载。

4. 测试

在客户端上测试nfs服务状况。

（1）在客户端上查看nfs服务器上的共享目录。

分别在ubuntu-chain0和ubuntu-chain1节点上查看。

在ubuntu-chain0节点上查看：

```
root@ubuntu-chain0:~# showmount --exports 192.168.47.132
Export list for 192.168.47.132:
/root/nfs-share/wwwroot 192.168.47.*
```

在ubuntu-chain1节点上查看：

```
root@ubuntu-chain1:~# showmount --exports 192.168.47.132
Export list for 192.168.47.132:
/root/nfs-share/wwwroot 192.168.47.*
```

从命令显示的结果可以看出，客户端访问正常。

（2）挂载目录测试

分别在ubuntu-chain0和ubuntu-chain1节点上挂载目录。

在ubuntu-chain0节点上挂载目录：

root@ubuntu-chain0:~# mount 192.168.47.132:/root/nfs-share/wwwroot /root/stack-lnmp/wwwroot

在ubuntu-chain1节点上挂载目录：

root@ubuntu-chain1:~# mount 192.168.47.132:/root/nfs-share/wwwroot /root/stack-lnmp/wwwroot

（3）创建文件测试

① 创建测试文件，在ubuntu-chain1节点上的wwwroot文件夹中创建测试文件test.php。

root@ubuntu-chain1:~/stack-lnmp/wwwroot# vi test.php

输入保存退出：

```
<?php
$aa=666;
echo 'hello '.'good luck to you!'.$aa."\n";
echo date('Y-m-d H:i:s',time())."\n";
echo "PHP版本:".phpversion();
?>
```

② 在ubuntu-chain0节点上查看test.php文件，在ubuntu-chain0节点上查看是否有test.php文件。

root@ubuntu-chain0:~/stack-lnmp# ls wwwroot
test.php

显示test.php文件内容：

```
root@ubuntu-chain0:~/stack-lnmp# cat wwwroot/test.php
<?php
$aa=666;
echo 'hello '.'good luck to you!'.$aa."\n";
echo date('Y-m-d H:i:s',time())."\n";
echo "PHP版本:".phpversion();
?>
```

③ 在nfs服务器上查看文件和内容。

```
root@blockchain:~/nfs-share/wwwroot# ls
test.php
root@blockchain:~/nfs-share/wwwroot# cat test.php
<?php
$aa=666;
echo 'hello '.'good luck to you!'.$aa."\n";
echo date('Y-m-d H:i:s',time())."\n";
echo "PHP版本:".phpversion();
?>
```

至此nfs服务测试成功。

5. 浏览器测试

在浏览器中测试通过挂载的网页文件，如图7-22所示。

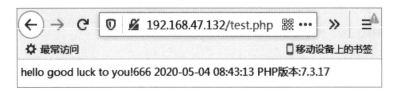

图 7-22　宿主机共享 nfs 服务器网页测试

从浏览器显示的结果可以看出，正是网页显示的内容，通过宿主机挂载nfs卷测试成功。

6. 实验完毕清除环境

完成整个测试，删除stack-lnmp服务栈。

```
root@blockchain:~# docker stack rm stack-lnmp
Removing service stack-lnmp_mysql
Removing service stack-lnmp_nginx
Removing service stack-lnmp_php
Removing network stack-lnmp_lnmp
```

五、网页卷直接挂载到 nfs 服务器上

上面的实验是通过本地的数据卷与nfs服务器的数据卷进行挂载，本地的数据卷再挂载到容器中实现，下面将通过nfs服务器上数据卷直接挂载到容器中实现。

为了区分该实验与前面的不同，先修改nfs数据卷上的网页内容，然后修改docker-compose.yml文件进行测试。

1. 修改原来的 test.php 文件内容

这里将"hello '.'good luck to you!"，修改成"******hello '.'good luck to you!******"进行区分。

```
root@blockchain:~/nfs-share/wwwroot# vi test.php
<?php
$aa=666;
echo '******hello '.'good luck to you!******'.$aa."\n";
echo date('Y-m-d H:i:s',time())."\n";
echo "PHP版本:".phpversion();
?>
```

2. 修改 docker-compose.yml 文件

这里重点是修改volumes部分，先定义volumes，然后在PHP和nginx服务中引用，具体内容在该文件中以蓝色的方式显示。关于docker-compose.yml文件中volumes定义可以参考官网：https://docs.docker.com/compose/compose-file/#volume-configuration-reference。

```
root@blockchain:~/stack-lnmp# vi docker-compose.yml
# 指定服务版本号
version: '3.8'
# 服务
services:
```

```yaml
# 服务名称
  mysql:
    # 指定服务容器名字
    hostname: mysql
    # 指定使用官方mysq 15.6版本
    image: mysql
    # 映射端口
    ports:
        - 3306:3306
    # 映射服务数据卷路径
    volumes:
        - ./mysql/conf:/etc/mysql/conf.d
        - ./mysql/data:/var/lib/mysql
    # 指定数据库变量
    environment:
    # 设置数据库密码，这里用123456，用户根据需要更改
        MYSQL_ROOT_PASSWORD: 123456
    # 添加user用户
        MYSQL_USER: user
    # 设置user用户密码
        MYSQL_PASSWORD: user123
    networks:
        - lnmp
    deploy:
      replicas: 2
      update_config:
        parallelism: 2
        delay: 10s
      restart_policy:
        condition: on-failure
# 服务名称
  php:
    # 指定服务容器名字
    hostname: php
    # 构建
    image: bitnami/php-fpm
      # 映射mysql服务别名
    ports:
          - 9000:9000
    # 映射服务数据卷路径
    volumes:
          - type: volume
            source: nfs-vol
            target: /usr/share/nginx/html
            volume:
```

```yaml
            nocopy: true
      networks:
          - lnmp
      stdin_open: true
      tty: true
      deploy:
        replicas: 2
        update_config:
          parallelism: 2
          delay: 10s
        restart_policy:
          condition: on-failure
# 服务名称
  nginx:
    # 指定服务容器名字
      hostname: nginx
    # 构建
      image: nginx
    # 映射数组级的端口
      ports:
          - 80:80
          - 443:443
    # 映射服务数据卷路径
      volumes:
          - ./nginx/nginx.conf:/etc/nginx/nginx.conf
          - type: volume
            source: nfs-vol
            target: /usr/share/nginx/html
            volume:
              nocopy: true
      networks:
          - lnmp
      deploy:
        replicas: 2
        update_config:
          parallelism: 2
          delay: 10s
        restart_policy:
          condition: on-failure
networks:
    lnmp:
      driver: overlay
volumes:
    nfs-vol:
      driver_opts:
```

```
type: "nfs"
o: "addr=192.168.47.132,nolock,soft,rw"
device: ":/root/nfs-share/wwwroot"
```

说明：

在文件的最后定义了volumes，nfs-vol是定义的卷的名字，用户可以根据实际需求定义具体的名字；driver_opts部分定义了类型"type: "nfs""，用来说明该卷是nfs类型；"o: "addr=192.168.47.132,nolock,soft,rw""定义了nfs服务器的地址，文件的读写权限，以及锁定；"device: ":/root/nfs-share/wwwroot""定义了具体的nfs服务器的共享目录。

服务中引用volumes部分，type: volume定义类型；source: nfs-vol定义源，即nfs服务器提供卷的名字，可以自己定义，但需要在顶级键volumes中定义；target:/usr/share/nginx/html部分定义容器中的目标目录，此处是nginx容器网页主目录的位置；volume: nocopy:true部分定义卷的复制属性。

3. 启动 stack 服务栈

```
root@blockchain:~/stack-lnmp# docker stack deploy -c docker-compose.yml stack-lnmp
Creating network stack-lnmp_lnmp
Creating service stack-lnmp_nginx
Creating service stack-lnmp_mysql
Creating service stack-lnmp_php
```

4. 测试

在浏览器中输入192.168.47.132/test.php测试，可以看到已经更改为更新过的nfs服务器页面，如图7-23所示。

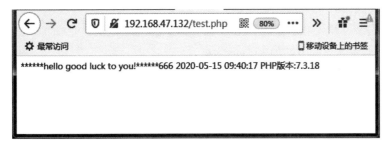

图 7-23　直接挂载 nfs 卷测试网页

至此，通过nfs服务挂载数据卷测试成功。

任务总结

本项目利用docker swarm中的stack强大功能部署lnmp网站，主要技术要点有：

① docker swarm stack配置文件与compose的配置文件格式基本一致，命令稍有差别，在compose配置文件中的命令在docker swarm stack配置文件不完全支持，可以参考官方网站查阅：https://docs.docker.com/compose/compose-file/。例如，本任务中的docker-compose.yml配置文件中将depend_on命令删除，docker swarm stack配置文件不支持该命令，通过启动顺序实现依赖关系。

② Nginx配置文件很重要，注意配置location部分，添加php解析参数。本任务中由于docker

swarm stack配置文件不支持depend_on和links命令，并没有使用php的别名，直接使用php名称。

③ 网页文件可以存放在第三方的服务器上，本任务测试了存放在nfs服务器上，通过挂载的方式提供网页服务。nfs服务需要nfs和rpc两个服务共同支持，因此可以将nfs和rpc服务设置成服务器自动启动方式。

任务扩展

本任务测试了运用docker-swarm部署lnmp网站服务，并通过nfs服务挂载网站的网页数据卷，读者尝试用该方法部署lamp网站进行测试。

任务四　运用 Docker Stack 部署高可用个人博客网站

任务场景

公司已经创建了多主机的环境，计划将前期通过compose部署的个人博客网站移植到现在的多主机环境，并且要求改善当前单一manager主机的模式，采用双manager主机模式解决manager主机的冗余备份。

任务描述

本任务学习在当前环境中加入一台manager主机，使当前环境有两台manager主机，起到备份作用，达到整个集群的高可用，在此基础上部署个人博客网站。

任务目标

◎ 能够根据需求在集群中添加manager主机
◎ 能够根据需求设置nfs服务共享
◎ 能够根据需求编写stack file文件
◎ 能够运用docker-swarm部署个人博客网站

任务实施

一、环境设计

1. 架构设计

采用图7-24所示架构，WordPress由两个容器组成，分别是WordPress和mysql。两个容器的命名也分别用WordPress和mysql，分别开启两个副本，运行在ubuntu-chain0和ubuntu-chain1节点上，为了增加管理的可靠性，将ubuntu-chain0节点提升为manager角色，这样ubuntu-chain0既可承担manager角色又承担worker角色，增加网络可靠性（不建议在一个网络中设置manager数量为偶数，一般采用3、5、7等奇数）。

数据存储采用第三方的nfs服务器，本任务中用manager主机作为nfs服务器，实际应用中可以是其他第三方主机，或者公有云上。

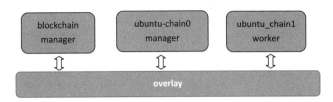

图 7-24　个人博客网站架构

2. 目录结构设计

（1）在nfs服务器上创建目录

在/root/nfs-share目录下创建db目录，存放数据库，目录结构如下：

```
root@blockchain:~/nfs-share# tree wordpress/
wordpress/
└── db
root@blockchain:~/nfs-share# mkdir wordpress
root@blockchain:~/nfs-share# cd wordpress/-
root@blockchain:~/nfs-share/wordpress# mkdir wp
```

（2）修改nfs配置文件

修改/etc/exports文件，按照如下内容修改。

```
/root/nfs-share 192.168.47.*(rw,fsid=0,sync,no_root_squash,no_subtree_check)
/root/nfs-share/wwwroot 192.168.47.*(rw,sync,no_root_squash,no_subtree_check)
/root/nfs-share/wordpress/db 192.168.47.*(rw,sync,no_root_squash,no_subtree_check)
```

其中，在/root/nfs-share文件夹一行添加了fsid=0参数，含义是共享目录的最上层父目录，wwwroot和wordpress/db的最上层父目录即为/root/nfs-share。fsid=0这个配置在/etc/exports中只能出现一次。

（3）重新启动nfs服务

```
root@blockchain:~# systemctl restart rpcbind
root@blockchain:~# systemctl restart nfs-kernel-server
```

（4）查看加载的目录

```
root@blockchain:~/nfs-share/wordpress# showmount --exports
Export list for blockchain:
/root/nfs-share/wordpress/db 192.168.47.*
/root/nfs-share/wwwroot      192.168.47.*
/root/nfs-share              192.168.47.*
```

从命令执行结果上看，/root/nfs-share/wordpress/db目录成功加载。

3. 测试

在客户端上测试nfs服务状况，用showmount命令分别在ubuntu-chain0和ubuntu-chain1节点上查看nfs服务器上的共享目录。

（1）在ubuntu-chain0节点上查看

```
root@ubuntu-chain0:~# showmount --exports 192.168.47.132
Export list for 192.168.47.132:
```

```
/root/nfs-share/wordpress/db 192.168.47.*
/root/nfs-share/wwwroot      192.168.47.*
/root/nfs-share              192.168.47.*
```

(2) 在ubuntu-chain1节点上查看

```
root@ubuntu-chain1:~# showmount --exports 192.168.47.132
Export list for 192.168.47.132:
/root/nfs-share/wordpress/db 192.168.47.*
/root/nfs-share/wwwroot      192.168.47.*
/root/nfs-share              192.168.47.*
```

从命令显示结果可以看出，客户端访问正常。

二、升级 ubuntu-chain0 节点

将ubuntu-chain0节点升级为manager节点。

1. 将节点离开集群

如果将ubuntu-chain0节点升级为manager，需要先将ubuntu-chain0节点离开集群，用docker swarm leave命令实现。

(1) 查看当前节点状况

```
root@blockchain:~/stack-wordpress# docker node ls
ID                            HOSTNAME         STATUS   AVAILABILITY   MANAGER STATUS   ENGINE VERSION
9paj3s5qmp3p8t7ogwokobmtu *   blockchain       Ready    Drain          Leader           19.03.8
6ote9w0lq1000twjm4ttl4765     ubuntu-chain0    Ready    Active                          19.03.8
rq166pl89u2ncoerhr0y3ltsr     ubuntu-chain1    Ready    Active                          19.03.8
```

可以看到有3个节点：blockchain、ubuntu-chain0和ubuntu-chain1，其中blockchain的管理状态是Leader。

(2) 将ubuntu-chain0节点离开集群

执行docker swarm leave命令。

```
root@ubuntu-chain0:~/wordpress# docker swarm leave
Node left the swarm.
```

(3) 再次查看节点状态

```
root@blockchain:~/stack-wordpress# docker node ls
ID                            HOSTNAME         STATUS   AVAILABILITY   MANAGER STATUS   ENGINE VERSION
9paj3s5qmp3p8t7ogwokobmtu *   blockchain       Ready    Drain          Leader           19.03.8
6ote9w0lq1000twjm4ttl4765     ubuntu-chain0    Down     Active                          19.03.8
rq166pl89u2ncoerhr0y3ltsr     ubuntu-chain1    Ready    Active                          19.03.8
```

从查看结果可以看出，ubuntu-chain0节点已经down了。

2. 升级 ubuntu-chain0 节点

docker swarm join-token [OPTIONS] (worker|manager)命令用来获取添加新的工作节点和管理

节点到Swarm的命令和Token。

（1）查看加入命令

在manager管理主机blockchain上运行docker swarm join-token manager命令，查看将升级ubuntu-chain0节点的命令格式。

```
root@blockchain:~/stack-wordpress# docker swarm join-token manager
To add a manager to this swarm, run the following command:
    docker swarm join --token SWMTKN-1-3hs612sa7oeijw7gb9vcrwa4x480f8hjrc0da5wvttzr7wd6uw-an010a50gbkq24vx10ykydk85 192.168.47.132:2377
```

复制该命令并在ubuntu-chain0节点上执行。

（2）执行docker swarm join --token命令

在ubuntu-chain0节点上执行加入管理角色的命令。

```
root@ubuntu-chain0:~/wordpress# docker swarm join --token SWMTKN-1-3hs612sa7oeijw7gb9vcrwa4x480f8hjrc0da5wvttzr7wd6uw-an010a50gbkq24vx10ykydk85 192.168.47.132:2377
This node joined a swarm as a manager.
```

从命令执行结果可以看出，ubuntu-chain0节点已经成为manager。

（3）验证

再次用docker node ls命令查看节点状况

```
root@blockchain:~/stack-wordpress# docker node ls
ID                            HOSTNAME         STATUS    AVAILABILITY   MANAGER STATUS    ENGINE VERSION
9paj3s5qmp3p8t7ogwokobmtu *   blockchain       Ready     Drain          Leader            19.03.8
styoh10g7gzbu0izb00u8le1m     ubuntu-chain0    Ready     Active         Reachable         19.03.8
rq166pl89u2ncoerhr0y3ltsr     ubuntu-chain1    Ready     Active                           19.03.8
```

从命令执行结果可以看出，ubuntu-chain0节点的MANAGER STATUS状态是Reachable，已经成为manager。

当前，环境中有两个manager，提高了网络的可靠性。swarm的管理节点内置有对HA的支持。即使一个或多个节点发生故障，剩余管理节点也会继续保证swarm的运转。Swarm实现了一种主从方式的多管理节点的HA，即使有多个管理节点，也总是仅有一个节点处于活动状态。

通常处于活动状态的管理节点称为"主节点"（leader），而主节点也是唯一一个会对swarm发送控制命令的节点。也就是说，只有主节点才会变更配置，或发送任务到工作节点。如果一个备用（非活动）管理节点接收到了swarm命令，则它会将其转发给主节点。

三、编辑配置文件

创建stack-wordpress目录并在该目录下编辑stack-wordpress.yml文件，重点修改了volumes部分，已经蓝色显示。

```
root@blockchain:~/stack-wordpress# vi stack-wordpress.yml
#指定版本
```

```yaml
version: '3.8'
#制定服务,本部分包含db和wordpress服务
services:
#    定义db容器
  db:
    image: mysql:5.7
    volumes:
            - type: volume
              source: db
              target: /var/lib/mysql
              volume:
                nocopy: true
ports:
    - 3306
    environment:
        MYSQL_ROOT_PASSWORD: wordpress
        MYSQL_DATABASE: wordpress
        MYSQL_USER: wordpress
        MYSQL_PASSWORD: wordpress
      #定义部署策略
deploy:
            replicas: 2
            update_config:
                parallelism: 2
                delay: 10s
            restart_policy:
                condition: on-failure
networks:
            - wp_network
#定义wordpress容器
  wordpress:
    image: wordpress
    ports:
        - "80:80"
    environment:
        WORDPRESS_DB_HOST: db:3306
        WORDPRESS_DB_USER: wordpress
        WORDPRESS_DB_PASSWORD: wordpress
        WORDPRESS_DB_NAME: wordpress

      #定义部署策略
deploy:
            replicas: 2
            update_config:
                parallelism: 2
                delay: 10s
            restart_policy:
```

```
                condition: on-failure
    networks:
              - wp_network
#声明网络
networks:
    wp_network:
        driver: overlay
#声明数据库挂载卷
volumes:
    db:
        driver_opts:
            type: "nfs"
            o: "addr=192.168.47.132,nolock,soft,rw"
            device: ":/root/nfs-share/wordpress/db"
```

四、启动和测试

1. 启动网站

(1) 启动网站

```
root@blockchain:~/stack-wordpress# docker stack deploy -c stack-wordpress.yml stack-wordpress
Creating network stack-wordpress_wp_network
Creating service stack-wordpress_db
Creating service stack-wordpress_wordpress
```

(2) 查看服务栈运行状况

```
root@blockchain:~/stack-wordpress# docker stack ps stack-wordpress
ID                    NAME                          IMAGE               NODE
DESIRED STATE         CURRENT STATE                 ERROR               PORTS
 784k5c5ay0pr          stack-wordpress_wordpress.1   wordpress:latest
ubuntu-chain1         Running                       Starting 3 seconds ago
 zqc23rlsrx9k          stack-wordpress_db.1          mysql:5.7
ubuntu-chain1         Running                       Running 13 seconds ago
 pguzvpycm2f0          stack-wordpress_wordpress.2   wordpress:latest
ubuntu-chain0         Running                       Starting 3 seconds ago
 qkw7wp9yccq0          stack-wordpress_db.2          mysql:5.7
ubuntu-chain0         Running                       Running 12 seconds ago
```

从命令执行结果可以看出，每个服务运行两个副本，分别运行于ubuntu-chain0和ubuntu-chain1节点上。

2. 测试

(1) 查看服务运行状况

```
root@blockchain:~/stack-wordpress# docker stack services stack-wordpress
ID              NAME                MODE           REPLICAS
IMAGE                               PORTS
 a5mkwrhog01a    stack-wordpress_db                 replicated    2/2
```

```
  mysql:5.7              *:30002->3306/tcp
    p4z4mx4ey4zg         stack-wordpress_wordpress    replicated       2/2
  wordpress:latest       *:8080->80/tcp
```

（2）查看nfs服务器中数据库内容

```
root@blockchain:~/nfs-share/wordpress/db# ls
  auto.cnf        ca.pem          client-key.pem     ibdata1      ib_logfile1    mysql
private_key.pem   server-cert.pem    sys
  ca-key.pem      client-cert.pem    ib_buffer_pool   ib_logfile0    ibtmp1
performance_schema   public_key.pem   server-key.pem    wordpress
```

从命令显示结果可以看出，原来的空文件夹已经有数据存在，数据库卷挂载正常。

（3）浏览器测试

在浏览器中输入192.168.47.132:8080宿主机的IP地址，显示欢迎界面，如图7-25所示，选择语言，单击Continue按钮，输入标示、用户名等信息，然后单击"安装WordPress"。

图7-25 个人博客访问成功界面

任务总结

本任务利用高可用的管理节点和数据库挂载的第三方nfs服务器上技术实现高可用的个人博客网站的搭建。

① 将其中的一个worker节点升级为manager节点，需要先将worker节点脱离集群，再将该节点升级到manager节点。

② 数据库卷挂载到第三方服务器上，可以提高冗余和性能，也可以存放在公有云上实现，读者可以自行尝试。

③ 本任务个人博客网站通过docker-stack实现了WordPress和mysql的冗余，并通过volumes方式将数据挂载到nfs服务器上，达到了高可用的目的。

任务扩展

建议进一步学习docker-swarm部署网络服务，练习stack file文件编写，尝试将本任务的数据卷挂载到公有云上测试。

项目八　使用 Kubernetes 编排网络服务

【项目综述】

根据安安公司的业务需求,网络服务编排和管理工作越来越重要,需要一个更加方便的管理平台管理,公司信息部门决定学习 Kubernetes,计划用 Kubernetes 作为编排工具。

【项目目标】

◎ 了解 Kubernetes 架构和组件的功能

◎ 能够安装部署 Kubernetes 组件

◎ 能够运用 Kubernetes 编排部署网络服务

◎ 能够运用 Kubernetes 管理部署网络服务

任务一　认识 Kubernetes 架构

任务场景

由于业务需求,安安公司已经将公司的网络服务业务迁移到容器上运行,现在计划通过 Kubernetes 进行统一管理和编排,作为信息技术部门,首先要认识和学习 Kubernetes 架构和组件,本次任务是要学习 Kubernetes 的架构和组件功能。

任务描述

本任务学习 kubernetes 架构和组件,学习 kubernetes 的关键概念,kubernetes 的架构,各个组件的功能,以及各个组件之间的关系。

任务目标

◎ 能够解释 Kubernetes 的关键概念

◎ 能够描述 Kubernetes 的架构

◎ 能够描述 Kubernetes 组件功能

◎ 能够描述 Kubernetes 组件之间的关系

任务实施

一、了解 Kubernetes

容器编排技术,在前面学习了 Compose、Swarm,它们都是 Docker 自己的编排系统。2017 年

10 月，Docker 宣布将在新版本中加入对 Kubernetes 的原生支持。

目前，AWS、Azure、Google、阿里云、腾讯云等主流公有云提供的是基于 Kubernetes 的容器服务；Rancher、CoreOS、IBM、Mirantis、Oracle、Red Hat、VMWare 等也在大力研发和推广基于 Kubernetes 的容器 CaaS 或 PaaS 产品。可以说，Kubernetes 是当前容器行业微服务架构中标准配置。

Kubernetes 作为微服务架构的管理平台有它天然的优势：

① Kubernetes 工作环境几乎没有限制。不管什么语言什么框架写的应用（Java、Python、Node.js），Kubernetes 都可以在物理服务器、虚拟机、云环境等环境中安全启动。

② Kubernetes 完全兼容各种云服务提供商，如 Google Cloud、Amazon、Microsoft Azure，还可以工作在 CloudStack、OpenStack、OVirt、Photon、VSphere等中。

③ Kubernetes 有先进的调度能力。Kubernetes 如果发现有节点工作不饱和，便会重新分配pod，高效地利用内存、处理器等资源。如果一个节点死机了，Kubernetes 会自动重新创建之前运行在此节点上的 pod，反之亦然，当负载降下来的时候，Kubernetes 也会自动缩减 pod 的数量。

④ Kubernetes 具有较强自动缩放能力。如果用户量突然暴增，现有的 pod 规模不足了，那么会自动创建出一批新的 pod，以适应当前需求。

总之，Kubernetes 具有负载均衡、健康检查、滚动升级、失败冗余、资源监控、日志访问、调试应用程序、认证和授权、容灾恢复、DevOps等功能。

下面逐步揭开kubernetes的面纱。

二、Kubernetes 的关键概念

1. cluster

cluster是计算、存储、网络资源的集合，由一组节点组成，这些节点可以是物理服务器或者虚拟机，之上安装了Kubernetes平台。

2. master

master是整个Cluster的大脑，用于控制 Kubernetes 所有节点，所有任务和资源的分配都由master统一调度管理。

3. node

除master之外的其他计算机称为 node（节点），可以是一台物理主机，也可以是虚拟机，是集群中的工作负载节点，会被Master分配一些工作负载（如docker容器）。当node死机时，其他工作负载会被转移到其他节点上执行。master节点也可以担任 node节点。

4. pod

pod 是Kubernetes 分配的最小工作单元，被部署在节点上，每个pod可以包含一个或多个容器。pod 中所有容器共享相同的网络空间和资源，即具有相同的 IP 地址、IPC、主机名称、卷及其他资源。

默认情况下，pod里面某个容器停止，Kubernetes会自动检测到并且重新启动这个pod，如果pod所在的节点死机，则会把这个节点上的所有pod重新调度到其他节点上。

5. service

service为一组 pod 提供单一稳定的名称和地址，pod可能经常会被销毁和重建，IP地址并不固定，用户不可能访问一个经常更换地址的服务，Kubernetes 用service代理后端的pod，统一暴露IP地址和端口，这样即使pod由于会被销毁和重建而更改IP地址，也不会影响用户访问。

6. namespace

Kubernetes 利用namespace实现资源的隔离。namespace 可以将一个物理的 cluster 逻辑上划分成多个虚拟 cluster，每个 cluster 都是一个 namespace。不同 namespace 里的资源是完全隔离的。

Kubernetes 默认创建了两个 namespace：defaul和kube-system。创建资源时如果不指定，将被放到defaul中；Kubernetes 自己创建的系统资源将放到kube-system中。

7. Controller Manager

Controller Manager 由 kube-controller-manager 和 cloud-controller-manager 组成，是 Kubernetes 的"大脑"，它通过 apiserver 监控整个集群的状态，并确保集群处于预期的工作状态。Controller Manager架构图如图8-1所示。（参考https://feisky.gitbooks.io/kubernetes/content/components/controller-manager.html）

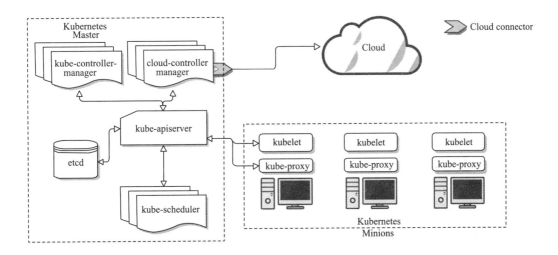

图 8-1　Controller Manager 架构图

（1）kube-controller-manager

kube-controller-manager 由一系列控制器组成，这些控制器可以划分为3组：

① 必须启动的控制器，包括：

- EndpointController；
- ReplicationController；
- PodGCController；
- ResourceQuotaController；
- NamespaceController；
- ServiceAccountController；
- GarbageCollectorController；
- DaemonSetController；
- JobController；
- DeploymentController；
- ReplicaSetController；
- HPAController；

- DisruptionController；
- StatefulSetController；
- CronJobController；
- CSRSigningController；
- CSRApprovingController；
- TTLController。

② 默认启动的可选控制器，可通过选项设置是否开启，包括：

- TokenController；
- NodeController；
- ServiceController；
- RouteController；
- PVBinderController；
- AttachDetachController。

③ 默认禁止的可选控制器，可通过选项设置是否开启，包括：

- BootstrapSignerController；
- TokenCleanerController。

(2) cloud-controller-manager

在 Kubernetes 启用 Cloud Provider 的时候才需要，用来配合云服务提供商的控制，也包括一系列的控制器，如：

- Node Controller；
- Route Controller；
- Service Controller。

三、认识 Kubernetes 架构

Kubernetes集群属于主从分布式架构，主要由一到多台Master Node和一到多台Worker Node，以及包括客户端kubectl命令行工具和其他附加选项构成。Kubernets整体架构如图8-2所示。

1. Master Node

Master Node作为控制节点，对集群进行整体调度管理，是整个集群的大脑，可以运行在物理机、虚拟机或者公有云上的虚拟机中。Master Node由API Server、Scheduler、Cluster State Store和Controller-Manger Server等组成，负责认证和授权、pod部署调度、扩容、状态存储、创建群集等工作。

(1) API server

API server是Kubernetes 的核心组件，提供其他模块之间的数据交互和通信的枢纽（其他模块通过 API server 查询或修改数据，只有 API server 才直接操作 etcd），支持 https（默认监听在6443 端口）和 http API（默认监听在 127.0.0.1 的 8080 端口）。其中 http API 是非安全接口，不做任何认证授权机制，不建议生产环境启用。

(2) Scheduler

Scheduler是kubernetes 核心组件，负责分配调度 Pod 到集群内的节点上。它监听 kube-apiserver，查询还未分配节点的pod，然后根据调度策略为这些 Pod 分配节点。请求到达后，由scheduler计算后端node的相关资源，如CPU、内存、容器运行数量等，然后选择合适的节点，由

项目八　使用 Kubernetes 编排网络服务

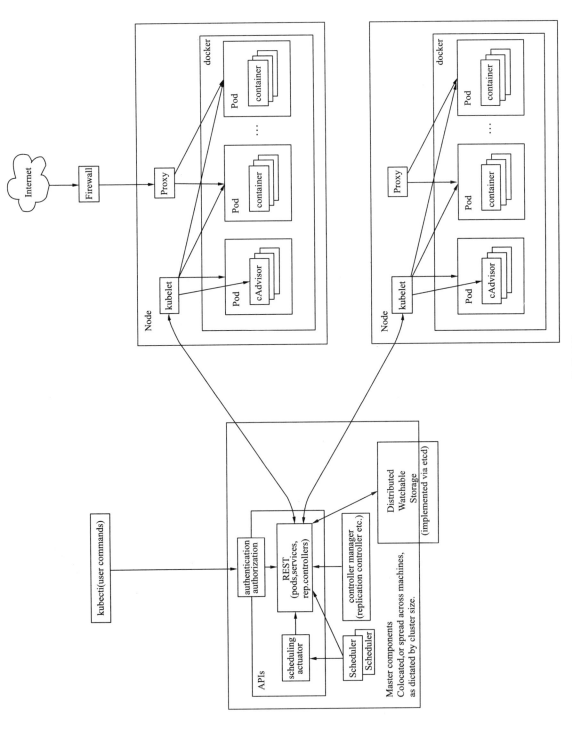

图 8-2　kubernetes 整体架构

节点的kubelet负责启动pod，即scheduler会先做预选，选择满足条件的节点，然后再做优选，选择最合适的node，启动pod。

（3）controller-manager

controller-manager是控制器管理器，负责统一管理集群中的控制器，维护集群的状态，如管控资源、健康检查、故障检测、自动扩展、滚动更新等，确保已创建的容器处于用户期望的状态运行。

由controller（控制器）具体对pod生命周期的管控，controller根据用户的需求启动和管理pod，可以在m个节点上运行n个pod，管理pod的健康性，使得pod能够按照指定的数量运行。每一组pod都需要独立的控制器来运行，实现跨节点自愈，管控pod的生命周期。控制器是通过label和label selector来实现感知自己管控组内的pod。

在Kubernetes环境中，不同的控制器满足客户不同类型的pod资源运行。常用控制器有如下几种：

① endpoint-controller：维护service和pod映射关系（关联信息由endpoint对象维护），保证service到pod的映射总是最新的。

② ReplicationController：定期关联replicationController和pod，保证replicationController定义的复制数量与实际运行pod的数量总是一致的，实现应用的扩容和缩容，控制pod的滚动更新和回滚，现逐步被ReplicaSetController代替。

③ DeploymentController：Deployment 为 pod 和 ReplicaSet 提供了一个声明式定义 (declarative) 方法，用来替代以前的 ReplicationController，来方便地管理应用。最常用的 Controller 大部分应用通过创建 Deployment 部署。Deployment 可以管理 pod 的多个副本，并确保 pod 按照期望的状态运行。

④ ReplicaSetController：ReplicaSet 与 ReplicationController 没有本质的不同，只是名字不同，并且 ReplicaSet 支持集合式的 selector（ReplicationController 仅支持等式）。虽然 ReplicaSet 可以独立使用，但建议使用 Deployment 来自动管理 ReplicaSet，这样就无须担心与其他机制的不兼容（如 ReplicaSet 不支持 rolling-update 但 Deployment 支持），并且可以支持版本记录、回滚、暂停升级等高级特性。

⑤ DaemonSetController：用于每个节点最多只运行一个 pod 副本的场景。常用来部署一些集群的日志、监控或者其他系统管理应用。典型的应用包括：

日志收集，如 fluentd、logstash 等。

系统监控，如 Prometheus Node Exporter、collectd、New Relic agent、Ganglia gmond 等。

系统程序，如 kube-proxy、kube-dns、glusterd, ceph 等。

⑥ StatefulSetController：为了解决有状态服务的问题（对应 Deployments 和 ReplicaSets 是为无状态服务而设计），支持稳定的持久化存储，即 pod 重新调度后还是能访问到相同的持久化数据，基于 PVC 来实现；稳定的网络标志，即 pod 重新调度后其 PodName 和 HostName 不变，基于 Headless Service（即没有 Cluster IP 的 Service）来实现；有序部署，有序扩展，即 pod 是有顺序的，在部署或者扩展的时候要依据定义的顺序依次依序进行（即从 $0 \sim N-1$，在下一个 pod 运行之前所有之前的 pod 必须都是 Running 和 Ready 状态），基于 init containers 来实现；有序收缩，有序删除（即从 $N-1 \sim 0$）。

⑦ JobController：用于运行结束就删除的应用。其他 Controller 中的 pod 通常是长期持续运行。

⑧ HPAController：Horizontal Pod Autoscaling（HPA）可以根据 CPU 使用率或应用自定

义 metrics 自动扩展 pod 数量（支持 replication controller、deployment 和 replica set）。控制器每隔 30 s（可以通过 --horizontal-pod-autoscaler-sync-period 修改）查询一次 metrics 的资源使用情况。

HPA支持3种 metrics 类型：预定义 metrics（如 Pod 的 CPU）以利用率的方式计算；自定义的 pod metrics，以原始值（raw value）的方式计算；自定义的 object metrics。

HPA支持两种 metrics 查询方式：Heapster 和自定义的 REST API。

HPA支持多 metrics。

⑨ ResourceQuotaController：资源配额（Resource Quotas）是用来限制用户资源用量的一种机制。它的工作原理为资源配额应用在 Namespace 上，并且每个 Namespace 最多只能有一个 ResourceQuota 对象。开启计算资源配额后，创建容器时必须配置计算资源请求或限制（也可以用 LimitRange 设置默认值）。用户超额后禁止创建新的资源。

⑩ ServiceAccountController：Service account 是为了方便 pod 里面的进程调用 Kubernetes API 或其他外部服务而设计的。它与 User account 不同。User account 是为人设计的，而 service account 则是为 pod 中的进程调用 Kubernetes API 而设计；User account 是跨 namespace 的，而 service account 则局限于它所在的 namespace；每个 namespace 都会自动创建一个 default service account。Token controller 检测 service account 的创建，并为它们创建 secret，开启 ServiceAccount Admission Controller 后，每个 pod 在创建后都会自动设置 spec.serviceAccountName 为 default（除非指定了其他 ServiceAccout），验证 pod 引用的 service account 已经存在，否则拒绝创建。如果 pod 没有指定 ImagePullSecrets，则把 service account 的 ImagePullSecrets 加到 pod 中，每个 container 启动后都会挂载该 service account 的 token 和 ca.crt 到 /var/run/secrets/kubernetes.io/serviceaccount/。

（4）etcd

etcd 是 CoreOS 基于 Raft 开发的分布式 key-value 存储，可用于服务发现、共享配置及一致性保障（如数据库选主、分布式锁等）。例如，apiserver对所有主机的操作结果，如创建pod、删除pod、调度pod的结果状态信息都保存在etcd中。如果这个插件异常，则整个集群运行都将异常，因为etcd异常后，整个集群的状态协议都将不能正常工作。因此etcd需要做高可用，以防止单点故障。

（5）label selector

label selector是标签选择器，根据标签选择符合条件资源的机制。其不仅仅用于pod资源，所有的对象都可以打上标签。service和controller都根据标签和标签控制器来识别pod资源。

2．worker node

节点上主要有3个组件：kubelet、kube-proxy和container engine。

（1）kubelet

相当于kubernetes的节点级的agent，每个节点上都运行一个 kubelet 服务进程，默认监听10250 端口，接收并执行 master 发来的指令，管理 pod 及 pod 中的容器。每个 kubelet 进程会在 API server 上注册所在节点的信息，定期向 master 节点汇报该节点的资源使用情况，并通过cAdvisor 监控节点和容器的资源。如当前节点的启动和当前节点状态监测和apiserver进行交互。

（2）kube-proxy

kube-proxy是kubernetes集群内部的负载均衡器。它是一个分布式代理服务器，在kubernetes的每个节点上都运行一个 kube-proxy 服务。它监听 API server 中 service 和 endpoint 的变化情况，并通过 iptables 等来为服务配置负载均衡（仅支持 TCP 和 UDP）。为当前节点的pod生成iptables或者ipvs规则，实现将用户请求调度到后端pod，为service组件服务，负责与apiserver随时保持通信，一旦发现某一service后的pod发生改变，需要将改变保存在apiserver中，apiserver内容发生改变后，会生成通知事件，使得所有关联apiserver的组件都能收到，而 kube-proxy可以收到这个通知事件，一旦发现某一service背后的pod发生改变，kube-proxy就会把改变反映在本地的iptables或者ipvs规则上，实现动态的变化。

kube-proxy 当前支持以下几种实现：

① userspace：最早的负载均衡方案，它在用户空间监听一个端口，所有服务通过 iptables 转发到这个端口，然后在其内部负载均衡到实际的 pod。该方式最主要的问题是效率低，有明显的性能瓶颈。

② iptables：目前推荐的方案，完全以 iptables 规则的方式来实现 service 负载均衡。该方式最主要的问题是在服务多的时候产生过多 iptables 规则，非增量式更新会引入一定的时延，大规模情况下有明显的性能问题。

③ ipvs：为解决 iptables 模式的性能问题，v1.11 新增了 ipvs 模式（v1.8 开始支持测试版），采用增量式更新，并可以保证 service 更新期间连接保持不断开。

注意：使用 ipvs 模式时，需要预先在每台 node 上加载内核模块，如 nf_conntrack_ipv4、ip_vs、ip_vs_rr、ip_vs_wrr、ip_vs_sh 等。

（3）container engine

容器运行时（Container Runtime）是 Kubernetes 最重要的组件之一，负责真正管理镜像和容器的生命周期。Kubelet 通过容器运行时接口（Container Runtime Interface，CRI）与容器运行时交互，以管理镜像和容器。

Container Runtime Interface（CRI）是 Kubernetes v1.5 引入的容器运行时接口，它将 Kubelet 与容器运行时解耦，将原来完全面向 pod 级别的内部接口拆分成面向 Sandbox 和 Container 的 gRPC 接口，并将镜像管理和容器管理分离到不同的服务。

支持的Container Runtime主要有：

① Docker：核心代码依然保留在 kubelet 内部（pkg/kubelet/dockershim），是最稳定和特性支持最好的运行时。

② OCI（Open Container Initiative，开放容器标准）：容器运行时。

③ PouchContainer：阿里巴巴开源的胖容器引擎。

④Frakti：支持 Kubernetes v1.6+，提供基于 hypervisor 和 docker 的混合运行时，适用于运行非可信应用，如多租户和 NFV 等场景。

⑤ Rktlet：支持 rkt 容器引擎。

⑥ Virtlet：Mirantis 开源的虚拟机容器引擎，直接管理 libvirt 虚拟机，镜像必须是 qcow2 格式。

⑦ Infranetes：直接管理 IaaS 平台虚拟机，如 GCE、AWS 等。

此外，node节点上还有addons（附件），如dns，可以动态变动dns解析内容，如service的名称改了，会自动触发dns的记录进行更改。

四、Kubernetes 组件之间的访问逻辑

Kubernetes的访问实际由两个体系组成，分别是部署系统和服务访问系统。部署系统解决用户通过管理工具实现对资源的管理和部署，使Kubernetes具有提供某种服务的功能，这个是

由kubelet实现；服务访问系统是Kubernetes解决用户通过什么方式访问Kubernetes部署的服务的，这个是由service实现的。

1. 部署系统各组件的逻辑流程

①客户端提交创建请求，可以通过API Server的Restful API，也可以使用kubectl命令行工具。支持的数据类型包括JSON和YAML。

② API Server处理用户请求，存储pod数据到etcd。

③调度器通过API Server查看未绑定的pod，尝试为pod分配主机。

④过滤主机（调度预选）：调度器用一组规则过滤掉不符合要求的主机。例如，Pod指定了所需要的资源量，那么可用资源比pod需要的资源量少的主机会被过滤掉。

⑤主机打分（调度优选）：对第一步筛选出的符合要求的主机进行打分，在主机打分阶段，调度器会考虑一些整体优化策略。例如，把Replication Controller的副本分布到不同的主机上，使用最低负载的主机等。

⑥选择主机：选择打分最高的主机，进行binding操作，将结果存储到etcd中。

⑦kubelet根据调度结果执行pod创建操作：绑定成功后，scheduler会调用APIServer的API在etcd中创建一个boundpod对象，描述在一个工作节点上绑定运行的所有pod信息。运行在每个工作节点上的kubelet也会定期与etcd同步boundpod信息，一旦发现应该在该工作节点上运行的boundpod对象没有更新，则调用Docker API创建并启动pod内的容器。

2. 服务访问系统各组件的逻辑流程

服务访问系统各组件的逻辑结构如图8-3所示。集群外部用户（互联网用户）一般通过负载均衡器访问，或者通过node节点的IP地址访问。具体流程如下：

①客户端通过互联网访问（该目标地址可能是某互联网提供商的负载均衡器IP地址+端口，或者公有云负载均衡器IP地址+端口）。

②负载均衡器访问Kubernetes工作节点的IP地址+端口（Node Port）。

③kubernetes将请求重定向到Service对象的Cluster IP服务端口上。

④Service对象将该请求转发到后端pod对象的IP和服务监听端口。

⑤pod响应对应的服务请求。

当集群内部访问时，集群内的pod对象可以直接请求Service的Cluster IP，从而获得服务。

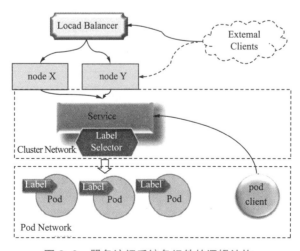

图8-3 服务访问系统各组件的逻辑结构

说明：

Service并不是一个具体的组件，而是一个通过规则定义出由多个pod对象组成的逻辑集合。Cluster IP是一个虚拟地址，并没有设备承载该IP，是由kube-proxy使用iptables规则重新定向的地址，专用于集群内通信。

Service的维护是由kube-proxy实现的。每个node上都运行着一个kube-proxy进程，kube-proxy是service的具体实现载体。

kube-proxy通过查询和监听API server中service和endpoint的变化，为每个service都建立了一个服务代理对象，并自动同步。服务代理对象是proxy程序内部的一种数据结构，它包括一个用于监听此服务请求的SocketServer，SocketServer的端口是随机选择的一个本地空闲端口。如果存在多个pod实例，kube-proxy同时也会负责负载均衡。而具体的负载均衡策略取决于Round Robin负载均衡算法及service的session会话保持这两个特性。会话保持策略使用的是ClientIP（将同一个ClientIP的请求转发同一个Endpoint上）。kube-proxy可以直接运行在物理机上，也可以以static-pod或者daemonset的方式运行。

Service是通过label selector实现与后端的多个Pod对象建立关联的。

五、Kubernetes 的网络架构

Kubernetes网络架构由3套体系组成，分别是node网络、service 网络、pod网络，架构如图8-4所示。

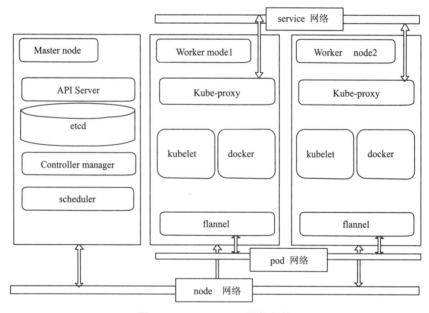

图 8-4　Kubernetes 网络架构

node网络本身不属于Kubernetes集群网络，属于组织内部的网络（部署在组织内部网络上的主机网络），或者公有云上主机实例所在的网络，目的是连接所有的master和worker主机节点的。

service 网络是kube-proxy通过iptables规则虚拟出来的ClientIP所在的网络，专用于Service资源对象。它是一个虚拟网络，用于为集群之中的Service配置IP地址，但是该地址不会配置在任何主机或容器的网络接口上，而是通过node上的kube-proxy配置为iptables或ipvs规则，从而将发往该地址的所有流量调度到后端的各pod对象之上。

pod网络是Kubernetes中的overlay网络，是不同node中pod通信所用的网络，专用于pod资源对象的虚拟网络，用于为各pod对象设定IP地址等网络参数，其地址配置在pod中容器的网络接口上。pod网络需要借助kubenet插件或CNI插件实现。

任务总结

Kubernetes是一个复杂的系统，重点要清楚如下几点：

① Kubernetes架构上属于分布式主从架构，主要分为master和worker节点，master节点实现接口和控制，worker节点是资源的实际提供者。

② Kubernetes各个组件可以直接运行在裸机、虚拟机和容器中，本任务采用容器方式运行。

③ etcd服务可以用专用节点运行，本任务中部署到master节点上运行。

④ Kubernetes部署系统由kubelet实现；服务访问由service实现。

⑤ Service不是一个具体的组件，而是一个通过规则定义出由多个pod对象组成的逻辑集合。

⑥ Cluster IP是一个虚拟地址，并没有设备承载该IP，而是由kube-proxy使用iptables规则重新定向的地址，专用于集群内通信。

⑦ Kubernetes网络架构由3套体系组成，分别是node网络、service 网络、pod网络。node网络解决节点之间的通信，service 网络解决Kubernetes为用户提供的服务，pod网络解决pod之间的通信。

任务扩展

参阅官方网站：https://kubernetes.io/docs/home/（英文）或者https://kubernetes.io/zh/docs/home/（中文）查阅相关概念和组件的功能，尝试自己画出各个组件的逻辑关系图。

任务二　部署和测试 Kubernetes 集群

任务场景

由于业务需求，安安公司已经将公司的网络服务业务迁移到容器上运行，现在决定通过Kubernetes进行统一管理和编排，本次任务是在测试环境中部署Kubernetes组件，并测试Kubernetes集群工作。

任务描述

本任务学习部署Kubernetes组件，创建Kubernetes集群，测试Kubernetes集群工作。

任务目标

◎ 能够实施准备部署Kubernetes部署的基本工作
◎ 能够安装部署Kubernetes的组件
◎ 能够创建Kubernetes集群，并加入工作节点
◎ 能够测试和查看Kubernetes集群工作状态

一、部署环境设计

采用前期的多主机模式创建的基础架构实施,如图8-5所示,blockchain主机作为master节点,ubuntu-chain0和ubuntu-chain1主机作为worker节点。

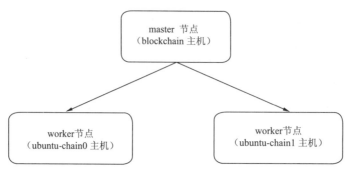

图 8-5 kubernetes 部署环境设计

二、部署 Kubernetes

安装 kubeadm、kubelet 和 kubectl 组件。

可参考官网参考:https://kubernetes.io/docs/setup/production-environment/container-runtimes/。

1. 安装准备

部署Kubernetes,需要在每台机器上安装以下的软件包:

kubeadm:用来初始化集群的指令,所有节点均安装。

kubelet:用来启动 pod 中的容器等,在集群中的每个节点上均安装。

kubectl:用来与集群通信的客户端命令行工具,例如,部署和管理应用,查看各种资源,创建、删除和更新各种组件。作为客户端的节点安装,可以所有节点均安装。

2. 安装和配置 docker

具体安装和配置步骤,参见本书项目一部分,此处不再赘述。安装和配置docker后,手动加载所有的配置文件,执行下面的命令。

```
root@blockchain:~# sysctl --system
```

3. 安装组件

(1) 安装基本工具

如果前期已经安装好apt-transport-https 和curl这些工具,此步可以省略。

```
root@blockchain:~# apt-get update && apt-get install -y apt-transport-https curl
   ...(此处省略了显示信息)
```

(2) 下载证书

```
root@blockchain:~# curl https://mirrors.aliyun.com/kubernetes/apt/doc/apt-key.gpg | apt-key add -
  % Total    % Received % Xferd  Average Speed   Time    Time     Time  Current
                                 Dload  Upload   Total   Spent    Left  Speed
```

```
100    653   100   653    0    0   4598    0 --:--:-- --:--:-- --:--:--  4598
OK
root@blockchain:~#        cat <<EOF >/etc/apt/sources.list.d/kubernetes.list
> deb https://mirrors.aliyun.com/kubernetes/apt/kubernetes-xenial main
> EOF
```

(3) 更新源列表

```
root@blockchain:~# apt-get update
… （此处省略了显示信息）
```

(4) 修改cgroupdriver

将cgroupdriver改成systemd，以使kubelet与docker兼容，这是由于kubelet默认的cgroupdriver是systemd，而docker默认的是cgroupfs（可以用命令docker info查看到）。

```
vi /etc/docker/daemon.json
{
    "registry-mirrors": ["https://y6akxxyg.mirror.aliyuncs.com"],
    "insecure-registries": ["192.168.47.132:5000"],
    "exec-opts": ["native.cgroupdriver=systemd"]     #增加一行
}
```

(5) 创建docker.service.d目录

```
root@blockchain:/etc/default# cd /etc/systemd/system/docker.service.d
```

(6) 重启Docker

```
root@blockchain:/# systemctl daemon-reload
root@blockchain:/# systemctl restart docker
```

(7) 安装Kubernetes组件

```
root@blockchain:/# apt-get install -y kubelet-1.18.2 kubeadm-1.18.2 kubectl-1.18.2
Reading package lists... Done
Building dependency tree
…  （此处省略了部分显示信息）
Setting up kubelet (1.18.3-00) ...
Setting up kubectl (1.18.3-00) ...
Setting up kubeadm (1.18.3-00) ...
root@blockchain:/#
```

以上操作是在master节点（blockchain主机）上实现的，另外两个节点也同样需要按照该方法安装Kubernetes组件。

三、初始化集群

1. 初始化 master

(1) 修改10-kubeadm.conf文件

修改10-kubeadm.conf配置文件，禁用swap（不禁用执行初始化命令会报错，不会成功）。增加环境变量Environment="KUBELET_SWAP_ARGS=---fail-swap-on=false"，然后在ExecStart=/usr/bin/kubelet中增加KUBELET_SWAP_ARGS参数。

```
vi /etc/systemd/system/kubelet.service.d/10-kubeadm.conf
…   （此处省略了部分显示信息）
EnvironmentFile=-/etc/default/kubelet
Environment="KUBELET_SWAP_ARGS=--fail-swap-on=false"
ExecStart=
ExecStart=/usr/bin/kubelet $KUBELET_KUBECONFIG_ARGS $KUBELET_CONFIG_ARGS
$KUBELET_KUBEADM_ARGS $KUBELET_EXTRA_ARGS $KUBELET_SWAP_ARGS
```

也可以用swapoff -a命令将swap禁掉。但这是临时禁掉，重启后不再生效。如果要永久禁用，可以修改/etc/fstab文件，将swap行注释掉。

（2）在master主机上用kubeadm init进行初始化

```
kubeadm init \
    --apiserver-advertise-address=192.168.47.132\
    --image-repository registry.aliyuncs.com/google_containers \
    --kubernetes-version=v1.18.2 \
    --pod-network-cidr=10.244.0.0/16 \
    --ignore-preflight-errors=Swap命令实现。
```

参数说明：

--apiserver-advertise-address

指明用 master 的哪个 interface 与 cluster 的其他节点通信。如果 master 有多个 interface，建议明确指定，如果不指定，kubeadm 会自动选择有默认网关的 interface。

--pod-network-cidr

指定 pod 网络的范围。Kubernetes 支持多种网络方案，而且不同网络方案对 --pod-network-cidr 有不同的要求，这里设置为 10.244.0.0/16，这是因为将使用 flannel 网络方案，必须设置成这个 CIDR。

--image-repository

Kubenetes默认Registries地址是 k8s.gcr.io，在国内并不能访问 gcr.io，在1.13版本中增加–image-repository参数，默认值是 k8s.gcr.io，将其指定为阿里云镜像地址：registry.aliyuncs.com/google_containers，这样可以下载指定的版本，作者使用最新版本letest下载没有成功，改为v1.18.2下载成功。

--kubernetes-version=v1.18.2

关闭版本探测，因为它的默认值是stable-1，会导致从https://dl.k8s.io/release/stable-1.txt下载最新的版本号，可以将其指定为固定版本（最新版：v1.18.1）来跳过网络请求。

--ignore-preflight-errors=Swap

忽略Swap参数错误。

下面是执行命令的输出（作者为了说明方便，将输出内容加了行号）：

```
root@blockchain:/# kubeadm init    --apiserver-advertise-address=
192.168.47.132    --image-repository registry.aliyuncs.com/google_containers
--kubernetes-version=v1.18.2    --pod-network-cidr=10.244.0.0/16
--ignore-preflight-errors=Swap
```

```
 1  W0527 02:09:40.725153    20500 configset.go:202] WARNING: kubeadm cannot
validate component configs for API groups [kubelet.config.k8s.io kubeproxy
    .config.k8s.io]
 2  [init] Using Kubernetes version: v1.18.2
 3  [preflight] Running pre-flight checks
 4          [WARNING Swap]: running with swap on is not supported. Please
disable swap
 5  [preflight] Pulling images required for setting up a Kubernetes cluster
 6  [preflight] This might take a minute or two, depending on the speed of
your internet connection
 7  [preflight] You can also perform this action in beforehand using 'kubeadm
config images pull'
 8  [kubelet-start] Writing kubelet environment file with flags to file "/var/
lib/kubelet/kubeadm-flags.env"
 9  [kubelet-start] Writing kubelet configuration to file "/var/lib/kubelet/
config.yaml"
10  [kubelet-start] Starting the kubelet
11  [certs] Using certificateDir folder "/etc/kubernetes/pki"
12  [certs] Generating "ca" certificate and key
13  [certs] Generating "apiserver" certificate and key
14  [certs] apiserver serving cert is signed for DNS names [blockchain
kubernetes kubernetes.default kubernetes.default.svc kubernetes.default.
svc.clu
        ster.local] and IPs [10.96.0.1 192.168.47.132]
15  [certs] Generating "apiserver-kubelet-client" certificate and key
16  [certs] Generating "front-proxy-ca" certificate and key
17  [certs] Generating "front-proxy-client" certificate and key
18  [certs] Generating "etcd/ca" certificate and key
19  [certs] Generating "etcd/server" certificate and key
20  [certs] etcd/server serving cert is signed for DNS names [blockchain
localhost] and IPs [192.168.47.132 127.0.0.1 ::1]
21  [certs] Generating "etcd/peer" certificate and key
22  [certs] etcd/peer serving cert is signed for DNS names [blockchain
localhost] and IPs [192.168.47.132 127.0.0.1 ::1]
23  [certs] Generating "etcd/healthcheck-client" certificate and key
24  [certs] Generating "apiserver-etcd-client" certificate and key
25  [certs] Generating "sa" key and public key
26  [kubeconfig] Using kubeconfig folder "/etc/kubernetes"
27  [kubeconfig] Writing "admin.conf" kubeconfig file
28  [kubeconfig] Writing "kubelet.conf" kubeconfig file
29  [kubeconfig] Writing "controller-manager.conf" kubeconfig file
30  [kubeconfig] Writing "scheduler.conf" kubeconfig file
31  [control-plane] Using manifest folder "/etc/kubernetes/manifests"
32  [control-plane] Creating static Pod manifest for "kube-apiserver"
33  [control-plane] Creating static Pod manifest for "kube-controller-
```

```
manager"
    34 W0527 02:09:59.440385   20500 manifests.go:225] the default kube-
apiserver authorization-mode is "Node,RBAC"; using "Node,RBAC"
    35 [control-plane] Creating static Pod manifest for "kube-scheduler"
    36 W0527 02:09:59.445236   20500 manifests.go:225] the default kube-
apiserver authorization-mode is "Node,RBAC"; using "Node,RBAC"
    37 [etcd] Creating static Pod manifest for local etcd in "/etc/kubernetes/
manifests"
    38 [wait-control-plane] Waiting for the kubelet to boot up the control
plane as static Pods from directory "/etc/kubernetes/manifests". This can
take up to 4m0s
    39 [apiclient] All control plane components are healthy after 38.515300
seconds
    40 [upload-config] Storing the configuration used in ConfigMap "kubeadm-
config"
in the "kube-system" Namespace
    41 [kubelet] Creating a ConfigMap "kubelet-config-1.18" in namespace kube-
system with the configuration for the kubelets in the cluster
    42 [upload-certs] Skipping phase. Please see --upload-certs
    43 [mark-control-plane] Marking the node blockchain as control-plane by
adding the label "node-role.kubernetes.io/master=''"
    44 [mark-control-plane] Marking the node blockchain as control-plane by
adding the taints [node-role.kubernetes.io/master:NoSchedule]
    45 [bootstrap-token] Using token: xm14qm.ha0pgl0utpn5tzwf
    46 [bootstrap-token] Configuring bootstrap tokens, cluster-info ConfigMap,
RBAC Roles
    47 [bootstrap-token] configured RBAC rules to allow Node Bootstrap tokens
to get nodes
    48 [bootstrap-token] configured RBAC rules to allow Node Bootstrap tokens
to post CSRs in order for nodes to get long term certificate credentials
    49 [bootstrap-token] configured RBAC rules to allow the csrapprover
controller automatically approve CSRs from a Node Bootstrap Token
    50 [bootstrap-token] configured RBAC rules to allow certificate rotation for
all node client certificates in the cluster
    51 [bootstrap-token] Creating the "cluster-info" ConfigMap in the "kube-
public" namespace
    52 [kubelet-finalize] Updating "/etc/kubernetes/kubelet.conf" to point to a
rotatable kubelet client certificate and key
    53 [kubelet-check] Initial timeout of 40s passed.
    54 [addons] Applied essential addon: CoreDNS
    55 [addons] Applied essential addon: kube-proxy
    56
    57 Your Kubernetes control-plane has initialized successfully!
    58
    59 To start using your cluster, you need to run the following as a regular
```

```
user:
 60
 61    mkdir -p $HOME/.kube
 62    sudo cp -i /etc/kubernetes/admin.conf $HOME/.kube/config
 63    sudo chown $(id -u):$(id -g) $HOME/.kube/config
 64
 65 You should now deploy a pod network to the cluster.
 66 Run "kubectl apply -f [podnetwork].yaml" with one of the options
listed at:
 67    https://kubernetes.io/docs/concepts/cluster-administration/addons/
 68
 69 Then you can join any number of worker nodes by running the following
on each as root:
 70
 71 kubeadm join 192.168.47.132:6443 --token xm14qm.ha0pgl0utpn5tzwf \
 72        --discovery-token-ca-cert-hash sha256:d87d1dfae264c2b5f3afb667adb1e
aadead8c93013ffad05caa9d8f4778c0676
```

命令输出说明（不同环境输出log信息不同，仅做参考）：

第2行，初始化，说明Kubernetes的版本是v1.18.2。

第3行，初始化前的检查。

第4行，说明要禁掉Swap。

第5~7行，说明下载所需要的镜像信息，用户可以提前下载，本任务已提前下载。

第8~10行，写入kubelet配置文件，启动kubelet。

第11~25行，初始化证书数据库。

第26~30行，利用/etc/kubernetes文件夹下的配置文件配置master节点的各个组件（如kubeadm、kubelete、controller-manager、scheduler）。

第31~44行，初始化master的组件，创建各个组件的pod，保存数据。

第45~52行，输出各个组件启动令牌认证信息。

第53行，kubelet-check检查时间。

第54~55行，安装附加组件CoreDNS和kube-proxy。

第57行，输出初始化成功信息。

第59~63行，提示用户怎样配置kubectl客户端。

第65~67行，提示怎样配置pod网络。

第69~72行，提示怎样将其他worker节点加入集群。

（3）配置kubectl客户端

可以按照第59~63行输出的提示信息配置kubectl客户端工具，配置完成后就可以用kubectl客户端工具对集群管理操作了。

说明：kubernetes不建议采用root用户进行管理（实际工作中都不建议直接用root用户进行管理），此处建议用普通用户。

切换至普通用户adminroot。

```
root@blockchain:~# su adminroot
```

```
adminroot@blockchain:/root$ cd ~
adminroot@blockchain:~$ pwd
/home/adminroot
```

执行第59~63行的提示信息。

```
adminroot@blockchain:~$ mkdir -p $HOME/.kube
[sudo] password for adminroot:
adminroot@blockchain:~$ sudo cp -i /etc/kubernetes/admin.conf $HOME/.kube/config
adminroot@blockchain:~$ sudo chown $(id -u):$(id -g) $HOME/.kube/config
```

执行完毕后，就可以用adminroot使用kubectl命令管理集群了，可以进行测试。

(4) 查看集群状况

① 查看集群状况，用kubectl get cs命令查看集群状况。

```
adminroot@blockchain:~$ kubectl get cs
NAME                 STATUS    MESSAGE             ERROR
scheduler            Healthy   ok
controller-manager   Healthy   ok
etcd-0               Healthy   {"health":"true"}
```

② 查看集群节点信息，用 kubectl get nodes命令查看集群节点信息，可以看到当前集群节点只有master节点，而且状态还是NotReady，说明还处于没准备好状态。

```
adminroot@blockchain:~$ kubectl get nodes
NAME         STATUS     ROLES    AGE    VERSION
blockchain   NotReady   master   101m   v1.18.3
```

③ 查看master节点详细信息，用kubectl describe node blockchain命令查看master节点（blockchain主机）详细信息。

```
adminroot@blockchain:~$ kubectl describe node blockchain
Name:               blockchain
Roles:              master
Labels:             beta.kubernetes.io/arch=amd64
                    beta.kubernetes.io/os=linux
                    kubernetes.io/arch=amd64
                    kubernetes.io/hostname=blockchain
                    kubernetes.io/os=linux
                    node-role.kubernetes.io/master=
Annotations:        kubeadm.alpha.kubernetes.io/cri-socket: /var/run/dockershim.sock
                    node.alpha.kubernetes.io/ttl: 0
                    volumes.kubernetes.io/controller-managed-attach-detach: true
CreationTimestamp:  Wed, 27 May 2020 02:10:34 +0000
Taints:             node-role.kubernetes.io/master:NoSchedule
                    node.kubernetes.io/not-ready:NoSchedule
```

```
  Unschedulable:       false
  Lease:
    HolderIdentity:    blockchain
    AcquireTime:       <unset>
    RenewTime:         Wed, 27 May 2020 03:54:36 +0000
  Conditions:
    Type               Status  LastHeartbeatTime              LastTransitionTime
  Reason                       Message
    ----               ------  -----------------              ------------------
    MemoryPressure     False   Wed, 27 May 2020 03:53:38 +0000    Wed, 27 May
  2020 02:10:24 +0000    KubeletHasSufficientMemory    kubelet has sufficient memory
  available
    DiskPressure       False   Wed, 27 May 2020 03:53:38 +0000    Wed, 27
  May 2020 02:10:24 +0000    KubeletHasNoDiskPressure     kubelet has no disk
  pressure
    PIDPressure        False   Wed, 27 May 2020 03:53:38 +0000    Wed, 27 May
  2020 02:10:24 +0000    KubeletHasSufficientPID       kubelet has sufficient PID
  available
    Ready              False   Wed, 27 May 2020 03:53:38 +0000    Wed, 27 May
  2020 02:10:24 +0000    KubeletNotReady               runtime network not
  ready: NetworkReady=false reason:NetworkPluginNotReady message:docker:
  network plugin is not ready: cni config uninitialized
  Addresses:
    InternalIP:  192.168.47.132
    Hostname:    blockchain
  Capacity:
    cpu:                4
    ephemeral-storage:  123327040Ki
    hugepages-1Gi:      0
    hugepages-2Mi:      0
    memory:             16402180Ki
    pods:               110
  Allocatable:
    cpu:                4
    ephemeral-storage:  113658199876
    hugepages-1Gi:      0
    hugepages-2Mi:      0
    memory:             16299780Ki
    pods:               110
  System Info:
    Machine ID:         42641c695ab74b388b9ac0c6ddb176aa
    System UUID:        1C044D56-6AF6-74AE-EC0C-DD93A758F5AD
    Boot ID:            633a0fb5-9118-46df-a9dd-34d461bde5c3
    Kernel Version:     4.15.0-101-generic
    OS Image:           Ubuntu 18.04.2 LTS
```

```
  Operating System:             linux
  Architecture:                 amd64
  Container Runtime Version:    docker://19.3.8
  Kubelet Version:              v1.18.3
  Kube-Proxy Version:           v1.18.3
PodCIDR:                        10.244.0.0/24
PodCIDRs:                       10.244.0.0/24
Non-terminated Pods:            (5 in total)
  Namespace         Name                                CPU Requests  CPU Limits   Memory Requests   Memory Limits   AGE
  ---------         ----                                ------------  ----------   ---------------   -------------   ---
  kube-system       etcd-blockchain                     0 (0%)        0 (0%)       0 (0%)            0 (0%)          103m
  kube-system       kube-apiserver-blockchain           250m (6%)     0 (0%)       0 (0%)            0 (0%)          103m
  kube-system       kube-controller-manager-blockchain  200m (5%)     0 (0%)       0 (0%)            0 (0%)          103m
  kube-system       kube-proxy-tp7vc                    0 (0%)        0 (0%)       0 (0%)            0 (0%)          103m
  kube-system       kube-scheduler-blockchain           100m (2%)     0 (0%)       0 (0%)            0 (0%)          103m
Allocated resources:
  (Total limits may be over 100 percent, i.e., overcommitted.)
  Resource           Requests    Limits
  --------           --------    ------
  cpu                550m (13%)  0 (0%)
  memory             0 (0%)      0 (0%)
  ephemeral-storage  0 (0%)      0 (0%)
  hugepages-1Gi      0 (0%)      0 (0%)
  hugepages-2Mi      0 (0%)      0 (0%)
Events:              <none>
```

通过 kubectl describe 命令的输出，可以看到节点的名字、角色、IP地址、系统配置、容器运行、资源配额等信息，也可以明显看出节点处于NodeNotReady状态，KubeletNotReady和runtime network not ready，表明尚未部署任何网络插件。

（5）配置pod网络

① 配置raw.githubusercontent.com地址解析，先修改hosts文档，加入一行"151.101.76.133 raw.githubusercontent.com"，目的是能正确解析raw.githubusercontent.com的IP地址151.101.76.133（该地址是我国香港地区的地址，全球有多个地址对应该域名的，151.101.76.133并不是唯一地址），互联网不能正确解析raw.githubusercontent.com地址的，可以用nslookup命令进行测试，如果能正确解析，该步骤可以省略。

```
adminroot@blockchain:~$ sudo vi /etc/hosts
[sudo] password for adminroot:
127.0.0.1       localhost.localdomain   localhost
```

```
::1             localhost6.localdomain6 localhost6
# The following lines are desirable for IPv6 capable hosts
::1     localhost ip6-localhost ip6-loopback
fe00::0 ip6-localnet
ff02::1 ip6-allnodes
ff02::2 ip6-allrouters
ff02::3 ip6-allhosts
151.101.76.133 raw.githubusercontent.com
```

② 配置pod网络，可以按照第65~67行输出的提示信息配置pod网络。

```
adminroot@blockchain:~$ kubectl apply -f https://raw.githubusercontent.com/coreos/flannel/master/Documentation/kube-flannel.yml
podsecuritypolicy.policy/psp.flannel.unprivileged created
clusterrole.rbac.authorization.k8s.io/flannel created
clusterrolebinding.rbac.authorization.k8s.io/flannel created
serviceaccount/flannel unchanged
configmap/kube-flannel-cfg configured
daemonset.apps/kube-flannel-ds-amd64 created
daemonset.apps/kube-flannel-ds-arm64 created
daemonset.apps/kube-flannel-ds-arm created
daemonset.apps/kube-flannel-ds-ppc64le created
daemonset.apps/kube-flannel-ds-s390x created
```

2. 将 worker node 接入集群

(1) 修改10-kubeadm.conf文件

修改worker node节点的10-kubeadm.conf配置文件，禁用swap，不禁用执行命令不会成功。增加环境变量Environment="KUBELET_SWAP_ARGS=--fail-swap-on=false"，然后在ExecStart=/usr/bin/kubelet中增加KUBELET_SWAP_ARGS参数。

```
root@ubuntu-chain0:~#vi /etc/systemd/system/kubelet.service.d/10-kubeadm.conf
  Environment="KUBELET_CONFIG_ARGS=--config=/var/lib/kubelet/config.yaml"
  # This is a file that "kubeadm init" and "kubeadm join" generates at runtime, populating the KUBELET_KUBEADM_ARGS variable
  dynamically
  EnvironmentFile=-/var/lib/kubelet/kubeadm-flags.env
  # This is a file that the user can use for overrides of the kubelet args as a last resort. Preferably, the user should use
  # the .NodeRegistration.KubeletExtraArgs object in the configuration files instead. KUBELET_EXTRA_ARGS should be sourced fr
  om this file.
  EnvironmentFile=-/etc/default/kubelet
  Environment="KUBELET_SWAP_ARGS=--fail-swap-on=false"
  ExecStart=
  ExecStart=/usr/bin/kubelet $KUBELET_KUBECONFIG_ARGS $KUBELET_CONFIG_ARGS $KUBELET_KUBEADM_ARGS $KUBELET_EXTRA_ARGS $KUBELET_SWAP_ARGSo
```

(2)安装kubernetes组件

略,可参见blockchain主机的操作。

(3)修改/etc/docker/daemon.json文件

将"native.cgroupdriver=systemd"加入到文件中(具体操作可参见blockchain主机的操作该部分的操作)。

(4)将worker节点接入集群

如果当时没记下来加入命令,可以用kubeadm token create --print-join-command命令在master节点上重新生成。

```
root@ubuntu-chain0:~# kubeadm join 192.168.47.132:6443 --ignore-preflight-errors=Swap --token xm14qm.ha0pgl0utpn5tzwf    --discovery-token-ca-cert-hash sha256:d87d1dfae264c2b5f3afb667adb1eaadead8c93013ffad05caa9d8f4778c0676
    W0527 10:13:08.833310   94691 join.go:346] [preflight] WARNING: JoinControlPane.controlPlane settings will be ignored when control-plane flag is not set.
    ...  (此处省略了部分显示信息)
Run 'kubectl get nodes' on the control-plane to see this node join the cluster.
```

(5)将ubuntu-chain1节点接入到集群

用同样的方法,将ubuntu-chain1节点接入到集群。

3. 查看和测试

(1)查看节点信息

用kubectl get nodes命令查看节点信息,可以看到3个节点都处于Ready状态,表明3个节点都已经加入到集群,并且处于可用状态(如果刚执行完加入节点命令初次查看3个节点没有全部处于Ready状态,可以稍等之后再次查看,系统可能还没有完全准备好)。

```
root@blockchain:~# kubectl get nodes
NAME              STATUS   ROLES    AGE   VERSION
blockchain        Ready    master   8h    v1.18.3
ubuntu-chain0     Ready    <none>   22m   v1.18.3
ubuntu-chain1     Ready    <none>   22m   v1.18.3
```

可以用kubectl get node -o wide命令查看详细信息。

```
adminroot@blockchain:~$ kubectl get node -o wide
NAME              STATUS   ROLES    AGE    VERSION   INTERNAL-IP      EXTERNAL-IP   OS-IMAGE             KERNEL-VERSION        CONTAINER-RUNTIME
blockchain        Ready    master   10h    v1.18.3   192.168.47.132   <none>        Ubuntu 18.04.2 LTS   4.15.0-101-generic    docker://19.3.8
ubuntu-chain0     Ready    <none>   121m   v1.18.3   192.168.47.137   <none>        Ubuntu 18.04.2 LTS   4.15.0-101-generic    docker://19.3.8
ubuntu-chain1     Ready    <none>   120m   v1.18.3   192.168.47.139   <none>        Ubuntu 18.04.2 LTS   4.15.0-101-generic    docker://19.3.8
```

(2)查看pod的命名空间信息

可以用 kubectl get pod --all-namespaces命令查看pod的命名空间,Kubernetes的组件均运行

在kube-system命名空间中,所有组件均处于Running状态(如果刚执行完加入节点命令初次查看3个节点的组件可能没有全部处于Running状态,可以稍等之后再次查看,系统需要下载一些镜像,可能还没有完全成功)。

```
root@blockchain:~# kubectl get pod --all-namespaces
NAMESPACE     NAME                                    READY   STATUS    RESTARTS   AGE
kube-system   coredns-7ff77c879f-jflgj                1/1     Running   0          8h
kube-system   coredns-7ff77c879f-w5rkg                1/1     Running   0          8h
kube-system   etcd-blockchain                         1/1     Running   0          8h
kube-system   kube-apiserver-blockchain               1/1     Running   0          8h
kube-system   kube-controller-manager-blockchain      1/1     Running   3          8h
kube-system   kube-flannel-ds-amd64-4gh58             1/1     Running   0          22m
kube-system   kube-flannel-ds-amd64-5wbnv             1/1     Running   0          22m
kube-system   kube-flannel-ds-amd64-m7g7h             1/1     Running   0          157m
kube-system   kube-proxy-mb8jf                        1/1     Running   0          22m
kube-system   kube-proxy-s4z8c                        1/1     Running   0          22m
kube-system   kube-proxy-tp7vc                        1/1     Running   0          8h
kube-system   kube-scheduler-blockchain               1/1     Running   5          8h
```

为该命令加上-o wide,可以看到更加详细的信息,如节点运行有哪些组件,以及在master节点(blockchain主机)上运行的coredns和etcd服务。

```
adminroot@blockchain:~$ kubectl get pods --all-namespaces -o wide
NAMESPACE     NAME                                    READY   STATUS    RESTARTS
   AGE         IP              NODE              NOMINATED NODE    READINESS GATES
  kube-system   coredns-7ff77c879f-jflgj               1/1     Running   0          10h
10.244.0.2      blockchain        <none>            <none>
  kube-system   coredns-7ff77c879f-w5rkg               1/1     Running   0          10h
10.244.0.3      blockchain        <none>            <none>
  kube-system   etcd-blockchain                        1/1     Running   0          10h
192.168.47.132  blockchain        <none>            <none>
  kube-system   kube-apiserver-blockchain              1/1     Running   0          10h
192.168.47.132  blockchain        <none>            <none>
  kube-system   kube-controller-manager-blockchain     1/1     Running   4          10h
192.168.47.132  blockchain        <none>            <none>
  kube-system   kube-flannel-ds-amd64-4gh58            1/1     Running   0          122m
192.168.47.139  ubuntu-chain1     <none>            <none>
  kube-system   kube-flannel-ds-amd64-5wbnv            1/1     Running   0          122m
192.168.47.137  ubuntu-chain0     <none>            <none>
  kube-system   kube-flannel-ds-amd64-m7g7h            1/1     Running   0          4h17m
192.168.47.132  blockchain        <none>            <none>
  kube-system   kube-proxy-mb8jf                       1/1     Running   0          122m
  192.168.47.137  ubuntu-chain0   <none>            <none>
  kube-system   kube-proxy-s4z8c                       1/1     Running   0          122m
192.168.47.139  ubuntu-chain1     <none>            <none>
  kube-system   kube-proxy-tp7vc                       1/1     Running   0          10h
```

192.168.47.132	blockchain		<none>		<none>
kube-system	kube-scheduler-blockchain		1/1	Running	6	10h
192.168.47.132	blockchain		<none>		<none>

至此，Kubernetes集群创建完成。从查看的信息中可以看到中master和worker节点运行正常，可以提供服务。

任务总结

Kubernetes支持多种runtime，可以采用Docker、CRI-O、Containerd、frakti等。runtime安装步骤，本任务采用Docker实现，需要所有主机首先安装好Docker。下面总结运用Docker runtime部署Kubernetes的要点。

① 部署Kubernetes时需要关闭swap，也就是禁止使用虚拟内存，因为使用硬盘进行虚拟内存交换影响速度，降低性能，Kubernetes为了提高性能，需要关闭swap。

② cgroupdriver驱动修改，可以将docker的cgroupdriver改成systemd，使kubelet与docker兼容，这是由于kubelet默认的cgroupdriver是systemd，而docker默认的是cgroupfs（可以用命令docker info查看到）；也可以将kubelet的cgroupdriver驱动修改为cgroupfs，只要二者兼容即可。

③ worker节点加入到集群后节点状态的改变需要一定的时间，因为需要下载一些镜像，具体时间这取决于用户网络速度；如果不成功，可以再次执行命令测试。

任务扩展

本任务通过虚拟机部署k8s集群，尝试在私有云环境中部署k8s集群，有条件的读者也可以尝试在公有云上通过实例部署该环境，并测试k8s集群。

任务三　通过NFS网络卷部署Kubernetes Nginx集群服务

任务场景

由于业务需求，安安公司已经完成部署Kubernetes集群，现测试Nginx集群服务，并能根据用户流量的需求进行伸缩处理，满足用户的需求。

任务描述

本任务学习编写编排Kubernetes集群服务的yaml文件，通过Kubernetes实现Nginx集群的创建，查看工作状态，然后根据用户流量的需求测试Nginx集群的伸缩，并测试在节点故障的情况下Kubernetes如何实现用户的期望。

任务目标

◎ 能够编写Kubernetes部署服务的yaml文件
◎ 能够运用Kubernetes部署Nginx服务

◎ 能够实现Nginx进行伸缩测试
◎ 能够查看和测试节点故障时Kubernetes处理方法

任务实施

一、架构设计

本任务仍采用前面图8-6所示的Kubernetes三个节点的架构环境，通过Kubernetes资源配置文件的方式部署Nginx集群，达到高可用的Web服务环境。

网页文件采用nfs服务器卷的方式直接提供pod，供pod中的容器使用。

本任务通过配置网页、部署Web服务、测试和查看结果、节点故障模拟、副本数量伸缩，学习kubernetes基本功能。

二、配置 nfs 网页文件

利用前期nfs配置的共享目录，如192.168.47.132上的/root/nfs-share/wwwroot共享文件夹。

```
adminroot@blockchain:~$ showmount --exports 192.168.47.132
Export list for 192.168.47.132:
/root/nfs-share/wordpress/db 192.168.47.*
/root/nfs-share/wwwroot      192.168.47.*
/root/nfs-share              192.168.47.*
```

下面创建一个测试网页。

```
adminroot@blockchain:~$ su
Password:
root@blockchain:/home/adminroot# cd /root/nfs-share/wwwroot
root@blockchain:~/nfs-share/wwwroot# ls
test.php
root@blockchain:~/nfs-share/wwwroot# echo "这是一个 kubernets nginx 集群测试页" > index.html
```

三、配置 kubernetes nginx 集群配置文件

1. 创建配置文件

创建kubernetes nginx集群配置文件文件夹

```
root@blockchain:/home/adminroot# mkdir k8s-nginx
root@blockchain:/home/adminroot# cd k8s-nginx/
root@blockchain:/home/adminroot/k8s-nginx#
```

2. 编写配置文件

下面编写配置文件，文件名命名web-nginx.yml，其格式仍采用命令格式。大部分命令已经在前期使用过或者含义比较明确，不再解释；对于新出现的或者本文件重点编辑的内容，在命令后用#进行解释。

```
root@blockchain:/home/adminroot/k8s-nginx# vi web-nginx.yml
apiVersion: apps/v1
kind: Deployment
```

```yaml
metadata:
  name: web-nginx
spec:
  selector:
    matchLabels:
      app: web-nginx
  replicas: 2          #副本数量
  template:
    metadata:
      labels:
        app: web-nginx
    spec:
      containers:
        - name: web
          image: nginx
          imagePullPolicy: Never   #如果已经有镜像，就不需要再拉取镜像
          ports:
            - name: web
              containerPort: 80
            # hostPort: 80      #将容器的80端口映射到宿主机的80端口
          volumeMounts:
            - name : nfs        #指定名称必须与下面一致
              mountPath: "/usr/share/nginx/html"   #容器内的挂载点
      volumes:
        - name: nfs             #指定名称必须与上面一致
          nfs:                  #nfs存储
            server: 192.168.47.132        #nfs服务器IP或域名
            path: "/root/nfs-share/wwwroot"    #nfs服务器共享的目录
```

在编辑过程中一定注意缩进的格式，如果有错误，查看缩进是否正确，然后修改，再尝试部署。

四、部署和测试

1. 部署 Nginx 集群

运用 kubectl apply 命令进行部署 Nginx 集群。

```
root@blockchain:/home/adminroot/k8s-nginx# kubectl apply -f web-nginx.yml
deployment.apps/web-nginx created
```

从命令执行结果可以看出，创建成功。

2. 查看 deployment 状态

运用 kubectl get deployment 命令查看当前 deployment 资源部署情况。

```
root@blockchain:/home/adminroot/k8s-nginx# kubectl get deployment
NAME        READY   UP-TO-DATE   AVAILABLE   AGE
web-nginx   2/2     2            2           8s
```

从命令执行结果可以看出，共有两个副本，两个有效副本，正是配置文件中定义的副本数量。

3. 查看 pod 运行情况

用kubectl get pod命令查看pod运行情况，本命令省去了NOMINATED NODE和READINESS GATES项。

```
root@blockchain:/home/adminroot/k8s-nginx# kubectl get pod -o wide
NAME                         READY   STATUS    RESTARTS   AGE   IP            NODE
web-nginx-66c8b4f869-bt55x   1/1     Running   0          15s   10.244.2.13   ubuntu-chain1
web-nginx-66c8b4f869-qtrwj   1/1     Running   0          15s   10.244.1.10   ubuntu-chain0
```

从命令执行结果可以看出，两个pod分别是web-nginx-66c8b4f869-bt55x和web-nginx-66c8b4f869-qtrwj，IP地址分别是10.244.2.13和10.244.1.10，分别分布在ubuntu-chain1和ubuntu-chain0节点。

4. 测试网页

用curl测试pod地址查看网页内容，执行的结果可以看出显示nfs上网页文件内容正确。

```
root@blockchain:/home/adminroot/k8s-nginx# curl 10.244.2.13
这是一个 kubernets nginx 集群测试页
root@blockchain:/home/adminroot/k8s-nginx# curl 10.244.1.10
这是一个 kubernets nginx 集群测试页
```

五、测试 kubernetes 的伸缩功能

1. 修改配置文件增加副本数量

修改web-nginx.yml文件，将replicas: 2改成4。

2. 重新执行部署命令

```
root@blockchain:/home/adminroot/k8s-nginx# kubectl apply -f web-nginx.yml
deployment.apps/web-nginx configured
```

3. 查看 deployment 状态

```
root@blockchain:/home/adminroot/k8s-nginx# kubectl get deployment
NAME        READY   UP-TO-DATE   AVAILABLE   AGE
web-nginx   4/4     4            4           11m
```

从命令执行结果可以看出，共有4个副本，4个有效副本，由原来的两个扩展成4个。

4. 查看 pod 运行情况

```
root@blockchain:/home/adminroot/k8s-nginx# kubectl get pod -o wide
NAME                         READY   STATUS    RESTARTS   AGE     IP            NODE            NOMINATED NODE   READINESS GATES
web-nginx-66c8b4f869-2fh8t   1/1     Running   0          2m23s   10.244.2.14   ubuntu-chain1   <none>           <none>
web-nginx-66c8b4f869-khb49   1/1     Running   0          12m     10.244.2.13   ubuntu-chain1   <none>           <none>
web-nginx-66c8b4f869-nl9qt   1/1     Running   0          2m23s   10.244.1.11   ubuntu-chain0   <none>           <none>
web-nginx-66c8b4f869-scnjm   1/1     Running   0          12m     10.244.1.10   ubuntu-chain0   <none>           <none>
```

从命令执行结果可以看出，在ubuntu-chain0和ubuntu-chain1上分别有两个pod，符合预期。

5. 测试网页

执行curl命令测试4个pod中容器的网页，显示正常。

```
root@blockchain:/home/adminroot/k8s-nginx# curl 10.244.1.10
这是一个 kubernets nginx 集群测试页
root@blockchain:/home/adminroot/k8s-nginx# curl 10.244.1.11
这是一个 kubernets nginx 集群测试页
root@blockchain:/home/adminroot/k8s-nginx# curl 10.244.2.13
这是一个 kubernets nginx 集群测试页
root@blockchain:/home/adminroot/k8s-nginx# curl 10.244.2.14
这是一个 kubernets nginx 集群测试页
```

六、测试 Kubernetes 的故障自愈功能

Kubernetes可以自动发现节点故障，会将不可用节点上的pod调度到可用的节点上，保持pod数量符合用户预期数量，下面测试以ubuntu-chain1故障为例查看Kubernetes自愈结果。

1. 将 ubuntu-chain1 主机关闭

```
root@ubuntu-chain1:~# halt -h
```

2. 查看节点状况

```
root@blockchain:/home/adminroot/k8s-nginx# kubectl get node
NAME             STATUS     ROLES    AGE     VERSION
blockchain       Ready      master   3h29m   v1.18.3
ubuntu-chain0    Ready      <none>   3h20m   v1.18.3
ubuntu-chain1    NotReady   <none>   3h20m   v1.18.3
```

可以看到ubuntu-chain1处于NotReady状态，不可提供服务。

3. 查看 deployment 状态

```
root@blockchain:/home/adminroot/k8s-nginx# kubectl get deployment
NAME        READY   UP-TO-DATE   AVAILABLE   AGE
web-nginx   2/4     4            2           24m
```

可以看到，web-nginx资源只有两个有效。

4. 查看 pod 状态

```
root@blockchain:/home/adminroot/k8s-nginx# kubectl get pod -o wide
  NAME                          READY   STATUS        RESTARTS   AGE   IP            NODE
NOMINATED NODE   READINESS GATES
  web-nginx-66c8b4f869-2fh8t    1/1     Terminating   0          18m   10.244.2.14   ubuntu-chain1   <none>           <none>
  web-nginx-66c8b4f869-2hwkl    1/1     Running       0          53s   10.244.1.13   ubuntu-chain0   <none>           <none>
  web-nginx-66c8b4f869-khb49    1/1     Terminating   0          28m   10.244.2.13   ubuntu-chain1   <none>           <none>
  web-nginx-66c8b4f869-nl9qt    1/1     Running       0          18m   10.244.1.11   ubuntu-chain0   <none>           <none>
```

```
   web-nginx-66c8b4f869-scnjm      1/1     Running    0      28m    10.244.1.10
ubuntu-chain0    <none>           <none>
   web-nginx-66c8b4f869-xcmfq      1/1     Running    0      53s    10.244.1.12
ubuntu-chain0    <none>           <none>
```

从命令执行结果可以看出，web-nginx-66c8b4f869-2fh8t和web-nginx-66c8b4f869-khb49两个pod都处于Terminating状态，共有4个pod处于Running状态，说明仍然保持4个pod副本运行，但这4个pod都是在ubuntu-chain0上运行。

5. 查看重新启动 ubuntu-chain1 主机后 deployment 和 pod 状态

（1）查看deployment状态

```
root@blockchain:/home/adminroot/k8s-nginx# kubectl get deployment
NAME         READY    UP-TO-DATE    AVAILABLE    AGE
web-nginx    4/4      4             4            39m
```

从命令执行结果可以看出，deployment状态恢复正常。

（2）查看pod状态

```
root@blockchain:/home/adminroot/k8s-nginx# kubectl get pod -o wide
   NAME        READY    STATUS       RESTARTS    AGE     IP       NODE          NOMINATED
NODE    READINESS GATES
   web-nginx-66c8b4f869-2hwkl      1/1     Running    0      12m    10.244.1.13
ubuntu-chain0    <none>           <none>
   web-nginx-66c8b4f869-nl9qt      1/1     Running    0      29m    10.244.1.11
ubuntu-chain0    <none>           <none>
   web-nginx-66c8b4f869-scnjm      1/1     Running    0      39m    10.244.1.10
ubuntu-chain0    <none>           <none>
   web-nginx-66c8b4f869-xcmfq      1/1     Running    0      12m    10.244.1.12
ubuntu-chain0    <none>           <none>
```

从命令执行结果可以看出，4个pod正常运行，但是仍在ubuntu-chain0上运行，并不会调度到ubuntu-chain1上运行。

（3）增加副本数量测试

将副本数量增加为5个测试。

```
root@blockchain:/home/adminroot/k8s-nginx# kubectl get pod -o wide
   NAME        READY    STATUS       RESTARTS    AGE     IP       NODE          NOMINATED
NODE    READINESS GATES
   web-nginx-66c8b4f869-2hwkl      1/1     Running    0      15m    10.244.1.13
ubuntu-chain0    <none>           <none>
   web-nginx-66c8b4f869-n9q4j      1/1     Running    0      8s     10.244.2.15
ubuntu-chain1    <none>           <none>
   web-nginx-66c8b4f869-nl9qt      1/1     Running    0      33m    10.244.1.11
ubuntu-chain0    <none>           <none>
   web-nginx-66c8b4f869-scnjm      1/1     Running    0      43m    10.244.1.10
ubuntu-chain0    <none>           <none>
   web-nginx-66c8b4f869-xcmfq      1/1     Running    0      15m    10.244.1.12
ubuntu-chain0    <none>           <none>
```

将副本数量增加为7个测试。

```
root@blockchain:/home/adminroot/k8s-nginx# kubectl get pod -o wide
NAME                          READY   STATUS    RESTARTS   AGE     IP            NODE             NOMINATED NODE   READINESS GATES
web-nginx-66c8b4f869-2hwkl    1/1     Running   0          19m     10.244.1.13   ubuntu-chain0    <none>           <none>
web-nginx-66c8b4f869-2hxss    1/1     Running   0          7s      10.244.2.16   ubuntu-chain1    <none>           <none>
web-nginx-66c8b4f869-9qlrd    1/1     Running   0          7s      10.244.2.17   ubuntu-chain1    <none>           <none>
web-nginx-66c8b4f869-n9q4j    1/1     Running   0          3m38s   10.244.2.15   ubuntu-chain1    <none>           <none>
web-nginx-66c8b4f869-nl9qt    1/1     Running   0          36m     10.244.1.11   ubuntu-chain0    <none>           <none>
web-nginx-66c8b4f869-scnjm    1/1     Running   0          46m     10.244.1.10   ubuntu-chain0    <none>           <none>
web-nginx-66c8b4f869-xcmfq    1/1     Running   0          19m     10.244.1.12   ubuntu-chain0    <none>           <none>
```

从命令执行结果可以看出，新增加的副本运行在ubuntu-chain1上，说明Kubernetes是可以将pod均衡调度到可用节点上的。

任务总结

① 采用Kubernetes资源配置文件方式部署应用，一定注意各个对象的缩进关系。

② api的版本可以通过kubectl api-versions查看，部署应用时选择自己需要的版本。

③ 本任务采用nfs实现卷的共享，直接将volume挂载到pod中，volumes定义是在template对象中的，这是持久卷共享其中一种方法，下一节将采用pv和pvc方式实现。

④ 本任务的伸缩采用手动更改replicas参数实现缩放功能，Kubernetes提供自动缩放功能称作Horizontal Pod Autoscaling (HPA)，是由HPAController实现。

⑤ 本方案中对网页的访问是通过内部的pod地址访问的，尚不能通过node节点的方式访问，是因为没有暴露到node节点端口，也没有采用service方式提供对外的服务，下一节将实践该功能。

任务扩展

本任务采用nfs实现卷的共享，读者可以在aws或者阿里云等公有云上申请存储，尝试公有云存储卷的挂载。

任务四 通过 PV 和 PVC 部署 Kubernetes Nginx 集群服务

任务场景

由于业务需求，安安公司已经完成部署Kubernetes集群，现测试Nginx集群的数据卷使用持久性的pv和pvc实现。

任务描述

本任务学习编写pv、pvc、service和Deployment配置文件，创建PV和PVC，部署Web服务，部署service资源，然后通过ClusterIP和node:port两种方式测试学习Kubernetes基本功能。

任务目标

◎ 能够描述Kubernetes的PV和PVC功能
◎ 能够编写Kubernetes编排需要的PV、PVC、service和Deployment的yaml配置文件
◎ 能够通过Kubernetes编排支持PV和PVC持久化数据卷的nginx服务
◎ 能够测试ClusterIP和node:port两种方式的nginx服务

任务实施

一、架构设计与功能需求分析

本任务仍采用前面图8-6所示的kubernetes三个节点的架构环境，通过Kubernetes资源配置文件的方式部署Nginx集群，达到高可用的Web服务环境。

网页文件使用Kubernetes的PV和PVC功能提供，在Kubernetes上创建PV，然后创建PVC，用来连接PV。PV使用nfs服务器上的共享文件夹实现。

网络服务的访问分别通过 ClusterIP和node:port两种方式实现。

本任务通过配置网页，编写PV、PVC、service和Deployment配置文件，创建PV和PVC，部署Web服务，部署service资源，然后通过ClusterIP和node:port两种方式测试学习Kubernetes基本功能。

二、认识 Kubernetes 卷

容器和 pod 可能会被频繁地销毁和创建。容器销毁时，保存在容器内部文件系统中的数据都会被清除。为了持久化保存容器的数据，可以使用 Kubernetes Volume实现。

Kubernetes Volume 是一个目录，当 Volume 被 mount 到 pod时，pod 中的所有容器都可以访问这个 Volume。Kubernetes Volume 也支持 emptyDir、hostPath、GCE Persistent Disk、AWS Elastic Block Store、NFS、Ceph 等。

1. emptyDir

emptyDir 是最基础的 Volume 类型，一个 emptyDir Volume 是 Host 上的一个空目录。emptyDir Volume 的生命周期与 pod 一致，是挂载到pod的，当 pod 从节点删除时，Volume 的内容也会被删除。但如果只是容器被销毁而 pod 还在，则 Volume 不受影响。emptyDir 特别适合 pod 中的容器需要临时共享存储空间的场景。

2. hostPath

hostPath Volume 是将 Docker Host 文件系统中已经存在的目录 mount 给 pod 的容器。hostPath 使用的场景是Kubernetes 或 Docker 内部数据（配置文件和二进制库）的应用。

3. 公有云上数据卷

公有云上部署Volume需要在公有云上创建和申请，需要将相关id和连接提供给Kubernetes。

4. Kubernetes 的 pv 和 pvc

Kubernetes 的 PersistentVolume 和 PersistentVolumeClaim解决方案可以让管理员创建存储卷提

供给用户使用。PersistentVolume (PV) 是外部存储系统中的一块存储空间, 由管理员创建和维护, 生命周期独立于 pod。PersistentVolumeClaim (PVC) 是对 PV 的申请 (Claim)。PVC 通常由普通用户创建和维护。需要为 pod 分配存储资源时, 用户可以创建一个 PVC, 指明存储资源的容量大小和访问模式（如只读）等信息, Kubernetes 会查找并提供满足条件的 PV 给用户使用。

PV是群集中的资源。PVC是对这些资源的请求, 并且还充当对资源的检查。PV和PVC之间的相互作用遵循以下生命周期：Provisioning → Binding → Using → Releasing → Recycling。

（1）供应准备Provisioning

通过集群外的存储系统或者云平台来提供存储持久化支持, 包含静态提供Static和动态提供Dynamic。

① 静态提供Static：集群管理员创建多个PV。它们携带可供集群用户使用的真实存储的详细信息。它们存在于Kubernetes API中, 可用于消费。

② 动态提供Dynamic：当管理员创建的静态PV都不匹配用户的PersistentVolumeClaim时, 集群可能会尝试为PVC动态配置卷。此配置基于StorageClasses, PVC必须请求一个类, 并且管理员必须已创建并配置该类才能进行动态配置。要求该类的声明有效地为自己禁用动态配置。

（2）绑定Binding

用户创建PVC并指定需要的资源和访问模式。在找到可用PV之前, PVC会保持未绑定状态。

（3）使用Using

用户可在pod中像volume一样使用PVC。

（4）释放Releasing

用户删除PVC来回收存储资源, PV将变成Released状态。由于还保留着之前的数据, 这些数据需要根据不同的策略来处理, 否则这些存储资源无法被其他PVC使用。

（5）回收Recycling

PV可以设置3种回收策略：保留（Retain）、回收（Recycle）和删除（Delete）。

① 保留策略：需要管理员手工回收数据。

② 删除策略：将删除PV和外部关联的存储资源, 需要插件支持。

③ 回收策略：将执行清除操作, 相当于执行rm -fr *操作。

三、编写配置文件

配置文件需要4个yaml配置文件, 一个用来创建PV, 一个用来创建PVC, 一个用来创建Nginx集群, 最后一个用来配置service。下面分别创建这4个yaml配置文件。

1. 创建 PV 配置文件

创建PV配置文件的目的是创建PV资源, 供用户通过PVC进行申请。本任务PV名称命名为web-pv, 空间大小定义为1 GB, 访问模式定义为ReadOnlyMany, PV回收策略定义为Retain, 定义存储class为nfs供PVC引用, 存储的位置为nfs服务器的 /root/nfs-share/wwwroot目录。

创建web-pv.yml文件, 输入以下内容：

```
root@blockchain:/home/adminroot/k8s-nginx# vi web-pv.yml
apiVersion: v1             #定义版本
kind: PersistentVolume     #声明资源类型
metadata:
    name: web-pv           #定义名称
spec:
```

```
      nfs:
        path: /root/nfs-share/wwwroot     #定义nfs路径
        server: 192.168.47.132            #定义nfs服务器的IP地址，也可用服务器名称
      #定义访问模式，可以是ReadWriteOnce或ReadWriteMany
      accessModes:
            - ReadOnlyMany
      capacity:
          storage: 1Gi           #定义PV的容量为1 GB
      #定义PV的分类为nfs，PVC可以指定class申请相应的class的PV
      storageClassName: nfs
          #定义PV回收策略
      persistentVolumeReclaimPolicy: Retain
```

配置文件中storage: 1Gi定义PV的容量为1 GB，当用户创建PVC申请时是不能大于1 GB空间的，也就是只能小于或者等于1 GB空间。

定义访问模式，可以是ReadWriteOnce、ReadWriteMany和ReadOnlyMany，ReadWriteOnce表示PV可以以ReadWrite模式mount到单个节点，ReadWriteMany表示PV可以以ReadWrite模式mount到多个节点，ReadOnlyMany可以以ReadOnly模式mount到多个节点。

2. 创建 PVC 配置文件

创建PVC配置文件的目的是创建PVC，用户可以通过PVC向PV申请存储资源。配置PVC的名称为web-pvc，以方便创建的应用引用，访问模式用ReadOnlyMany，申请的容量大小为1 GB，定义storageClass为nfs，即向storageClass为nfs的PV申请存储。

```
root@blockchain:/home/adminroot/k8s-nginx# vi web-pvc.yml
apiVersion: v1
kind: PersistentVolumeClaim
metadata:
   name: web-pvc
spec:
   accessModes:
         - ReadOnlyMany
   resources:
      requests:
          storage: 1Gi
   storageClassName: nfs
```

3. 配置 Nginx 应用配置文件

配置Nginx应用，将副本设置为4，也可以根据需要设置。该文件重点是在volumes中的设置，引用PVC数据，设置PVC提供数据，其claimName的值设置为所创建的PVC的名称web-pvc。

```
root@blockchain:/home/adminroot/k8s-nginx# vi web-nginx.yml
apiVersion: apps/v1
kind: Deployment
metadata:
   name: web-nginx
spec:
```

```
      selector:
        matchLabels:
          app: web-nginx
      replicas: 4
      template:
        metadata:
          labels:
            app: web-nginx
        spec:
          containers:
          - name: web
            image: nginx
            imagePullPolicy: Never
            ports:
            - name: web
              containerPort: 80
            volumeMounts:
            - name : nfs              #指定名称必须与下面一致
              mountPath: "/usr/share/nginx/html"   #容器内的挂载点
          volumes:
          - name: nfs                 #指定名称必须与上面一致
            persistentVolumeClaim:    #引用persistentVolumeClaim
              claimName: web-pvc
```

4. 创建 service 资源配置文件

Kubernetes提供服务是通过service实现的，service代表多个pod对象组成的逻辑集合，通过Cluster IP提供服务。Service的维护是由每个node上都运行kube-proxy实现的。通过创建service资源实现统一提供服务的功能。

编辑service资源配置文件，这里名称定义为web-nginx，用户可以根据实际情况定义。Selector标签选择器用web-nginx，这个要与nginx的labels定义要一致，说明哪些 label 的 pod 作为service 的后端。

```
root@blockchain:/home/adminroot/k8s-nginx# vi web-nginx-service.yml
apiVersion: v1
kind: Service   # 指明资源类型是 service
metadata:
  name: web-nginx  # service 的名字是 web-nginx
  labels:
     name: web-nginx   #服务的标签定义为web-nginx
spec:
    ports:  # 将 service 80 端口映射到 pod 的 80 端口，使用 TCP 协议
    - port: 80
      targetPort: 80
      protocol: TCP
    selector:
```

```
        app: web-nginx    # 指明哪些 label 的 pod 作为 service 的后端
```

四、创建资源和测试

本部分首先部署PV资源，然后通过PVC申请存储空间，再通过部署Nginx应用使用PVC申请到的资源，这样就可以通过访问Nginx的服务测试PV上的资源了，最后部署service资源，测试通过ClusterIP和node:port两种方式访问Nginx集群提供的Web服务。

1. 部署和查看 PV 资源

（1）部署PV资源

运用kubectl apply命令部署PV资源，用-f选项指定PV的配置文件。

```
root@blockchain:/home/adminroot/k8s-nginx# kubectl apply -f web-pv.yml
persistentvolume/web-pv created
```

从命令执行结果可以看出，web-pv已创建成功。下面查看web-pv资源。

（2）查看PV资源

用 kubectl get pv命令查看当前的PV状况。

```
root@blockchain:/home/adminroot/k8s-nginx# kubectl get pv
NAME      CAPACITY    ACCESS MODES    RECLAIM POLICY    STATUS      CLAIM   STORAGECLASS    REASON    AGE
web-pv    1Gi         ROX             Retain            Available           nfs                       119s
```

从命令执行结果可以看出，与配置文件定义的结果一致，PV的名称为web-pv，存储大小为1 GB，回收方式为Retain，当前处于有效状态，存储的位置为nfs。

（3）查看web-pv详细信息

用kubectl describe pv命令查看web-pv的详细信息。

```
root@blockchain:/home/adminroot/k8s-nginx# kubectl describe pv web-pv
Name:              web-pv
Labels:            <none>
Annotations:       Finalizers:    [kubernetes.io/pv-protection]
StorageClass:      nfs
Status:            Available
Claim:
Reclaim Policy:    Retain
Access Modes:      ROX
VolumeMode:        Filesystem
Capacity:          1Gi
Node Affinity:     <none>
Message:
Source:
    Type:      NFS (an NFS mount that lasts the lifetime of a pod)
    Server:    192.168.47.132
    Path:      /root/nfs-share/wwwroot
    ReadOnly:  false
Events:        <none>
```

web-pv详细信息除了名称、空间大小、状态、回收方式、文件系统、访问方式外,还详细显示nfs数据源信息。

2. 部署和查看 PVC 资源

(1) 部署PVC资源

运用kubectl apply命令部署PVC资源,用-f选项指定PVC的配置文件。

```
root@blockchain:/home/adminroot/k8s-nginx# kubectl apply -f web-pvc.yml
persistentvolumeclaim/web-pvc created
```

命令成功执行。

(2) 查看PVC资源

用 kubectl get pvc命令查看当前的PVC状况。

```
root@blockchain:/home/adminroot/k8s-nginx# kubectl get pvc
NAME      STATUS   VOLUME   CAPACITY   ACCESS MODES   STORAGECLASS   AGE
web-pvc   Bound    web-pv   1Gi        ROX            nfs            69s
```

从命令执行结果可以看出,与配置文件的定义的结果一致,PVC的名称为web-pvc,存储大小为1 GB,申请的PV为web-pv,当前处于Bound状态,存储STORAGECLASS为nfs。

(3) 查看web-pvc详细信息

用kubectl describe pvc命令显示web-pvc的详细信息。

```
root@blockchain:/home/adminroot/k8s-nginx# kubectl describe pvc web-pvc
Name:          web-pvc
Namespace:     default
StorageClass:  nfs
Status:        Bound
Volume:        web-pv
Labels:        <none>
Annotations:   pv.kubernetes.io/bind-completed: yes
               pv.kubernetes.io/bound-by-controller: yes
Finalizers:    [kubernetes.io/pvc-protection]
Capacity:      1Gi
Access Modes:  ROX
VolumeMode:    Filesystem
Mounted By:    <none>
Events:        <none>
```

3. 部署 Nginx 集群服务并测试

部署Nginx服务,该服务通过Deployment方式部署,网页文件引用PVC从PV申请的文件,并mount到容器的主页目录中。

(1) 部署Nginx集群服务

用kubectl apply命令,加上 -f指定配置文件 web-nginx.yml实现。

```
root@blockchain:/home/adminroot/k8s-nginx# kubectl apply -f web-nginx.yml
deployment.apps/web-nginx created
```

命令成功执行。

(2) 查看Deployment状况

```
root@blockchain:/home/adminroot/k8s-nginx# kubectl get deployment
NAME          READY   UP-TO-DATE   AVAILABLE   AGE
web-nginx     4/4     4            4           3m23s
```

可以看到4个pod副本在运行。

(3) 查看pod信息

```
root@blockchain:/home/adminroot/k8s-nginx# kubectl get pod
NAME                          READY   STATUS    RESTARTS   AGE
web-nginx-8657dfdc49-5l4kd    1/1     Running   0          4m29s
web-nginx-8657dfdc49-9c8fx    1/1     Running   0          4m29s
web-nginx-8657dfdc49-msh6z    1/1     Running   0          4m29s
web-nginx-8657dfdc49-qrjjv    1/1     Running   0          4m29s
```

下面查看pod详细信息。

```
root@blockchain:/home/adminroot/k8s-nginx# kubectl get pod -o wide
NAME                          READY   STATUS    RESTARTS   AGE     IP           NODE             NOMINATED NODE   READINESS GATES
  web-nginx-8657dfdc49-5l4kd  1/1     Running   0          4m42s   10.244.1.15  ubuntu-chain0    <none>           <none>
  web-nginx-8657dfdc49-9c8fx  1/1     Running   0          4m42s   10.244.1.16  ubuntu-chain0    <none>           <none>
  web-nginx-8657dfdc49-msh6z  1/1     Running   0          4m42s   10.244.2.4   ubuntu-chain1    <none>           <none>
  web-nginx-8657dfdc49-qrjjv  1/1     Running   0          4m42s   10.244.2.3   ubuntu-chain1    <none>           <none>
```

可以看到4个pod正在运行，IP地址分别是10.244.2.3、10.244.2.4、10.244.2.15、10.244.2.16，分别分布在ubuntu-chain0和ubuntu-chain1节点上，一切正常。

(4) 查看网页信息

```
root@blockchain:/home/adminroot/k8s-nginx# curl 10.244.2.3
这是一个 kubernets nginx 集群测试页
root@blockchain:/home/adminroot/k8s-nginx# curl 10.244.2.4
这是一个 kubernets nginx 集群测试页
root@blockchain:/home/adminroot/k8s-nginx# curl 10.244.1.16
这是一个 kubernets nginx 集群测试页
root@blockchain:/home/adminroot/k8s-nginx# curl 10.244.1.15
这是一个 kubernets nginx 集群测试页
```

用curl命令显示4个pod显示网页信息正常，说明pod中的容器确实通过PVC向PV申请到了存储，并mount到自己的主页文件夹中。

(5) 测试通过物理节点访问

```
root@blockchain:/home/adminroot/k8s-nginx# curl 192.168.47.132
curl: (7) Failed to connect to 192.168.47.132 port 80: Connection refused
root@blockchain:/home/adminroot/k8s-nginx# curl 192.168.47.137
```

```
curl: (7) Failed to connect to 192.168.47.137 port 80: Connection refused
root@blockchain:/home/adminroot/k8s-nginx# curl 192.168.47.139
curl: (7) Failed to connect to 192.168.47.139 port 80: Connection refused
```

从命令执行结果可以看出，通过物理节点方式访问是不成功的。下面通过service方式进行测试。

五、部署 service 并测试

本部分首先部署service，然后测试ClusterIP和node:port两种访问方式。

1. 部署 service 资源

```
root@blockchain:/home/adminroot/k8s-nginx# kubectl apply -f web-nginx-service.yml
service/web-nginx created
```

命令成功执行。

2. 查看 service 资源

```
root@blockchain:/home/adminroot/k8s-nginx# kubectl get service
NAME         TYPE        CLUSTER-IP     EXTERNAL-IP   PORT(S)   AGE
kubernetes   ClusterIP   10.96.0.1      <none>        443/TCP   26h
web-nginx    ClusterIP   10.101.16.24   <none>        80/TCP    49s
```

命令显示有两个ClusterIP：一个是kubernetes的，是kubernetes的Cluster内部通过这个service访问kubernetes API server的；另一个是web-nginx，该IP地址是10.101.16.24。注意，显示信息中web-nginx的type是ClusterIP，说明可以通过ClusterIP方式访问该service，下面会通过网页测试进行印证。

查看详细信息：

```
root@blockchain:/home/adminroot/k8s-nginx# kubectl describe service web-nginx
Name:              web-nginx
Namespace:         default
Labels:            name=web-nginx
Annotations:       Selector:  app=web-nginx
Type:              ClusterIP
IP:                10.101.16.24
Port:              <unset>  80/TCP
TargetPort:        80/TCP
Endpoints:         10.244.1.15:80,10.244.1.16:80,10.244.2.3:80 + 1 more...
Session Affinity:  None
Events:            <none>
```

详细信息中显示Endpoints，指明该ClusterIP实际是由10.244.1.15:80、10.244.1.16:80、10.244.2.3:80等提供，下面测试通过10.101.16.24访问网页。

3. 通过 ClusterIP 网页测试

```
root@blockchain:/home/adminroot/k8s-nginx# curl 10.101.16.24
这是一个 kubernets nginx 集群测试页
```

可以看到网页测试正常，那么用pod地址能访问吗？当然仍然是可以的，例如：

```
root@blockchain:/home/adminroot/k8s-nginx# curl 10.244.2.3
这是一个 kubernets nginx 集群测试页
```

4. 通过 node:port 网页测试

首先修改service的配置文件，添加type模式为node，然后进行网页测试。

(1) 修改service配置文件

在spec键中添加一行type: NodePort，然后保存退出。

```
root@blockchain:/home/adminroot/k8s-nginx# vi web-nginx-service.yml
apiVersion: v1
kind: Service    # 指明资源类型是 service
metadata:
  name: web-nginx   # service 的名字是 httpd-svc
  labels:
      name: web-nginx
spec:
    type: NodePort
    ports:  # 将 service 80 端口映射到 pod 的 80 端口，使用 TCP 协议
      - port: 80
        targetPort: 80
        protocol: TCP
    selector:
        app: web-nginx   # 指明哪些 label 的 pod 作为 service 的后端
```

(2) 重新部署service

```
root@blockchain:/home/adminroot/k8s-nginx# kubectl apply -f web-nginx-service.yml
service/web-nginx configured
```

命令执行成功。

(3) 查看service资源

```
root@blockchain:/home/adminroot/k8s-nginx# kubectl get service
NAME         TYPE        CLUSTER-IP      EXTERNAL-IP   PORT(S)         AGE
kubernetes   ClusterIP   10.96.0.1       <none>        443/TCP         27h
web-nginx    NodePort    10.101.16.24    <none>        80:32028/TCP    60m
```

看到web-nginx的type是NodePort，不再是ClusterIP，而且PORT(S)信息显示为80:32028/TCP，说明映射物理主机的端口是32028，下面进行网页测试。

(4) 通过物理主机的端口测试网页

```
root@blockchain:/home/adminroot/k8s-nginx# curl 192.168.47.132:32028
这是一个 kubernets nginx 集群测试页
root@blockchain:/home/adminroot/k8s-nginx# curl 192.168.47.137:32028
这是一个 kubernets nginx 集群测试页
root@blockchain:/home/adminroot/k8s-nginx# curl 192.168.47.139:32028
这是一个 kubernets nginx 集群测试页
```

通过对blockchain、ubuntu-chain0和ubuntu-chain1节点上网页测试，网页显示正常，不论是集群中的那个物理节点均可以访问Nginx集群提供的服务。那么通过ClusterIP地址能访问吗？当然也是可以访问的，测试如下：

```
root@blockchain:/home/adminroot/k8s-nginx# curl 10.101.16.24
这是一个 kubernets nginx 集群测试页
```

至此，测试完成，实现了通过PV和PVC提供应用卷的支持，实现了持久化数据保存，并通过service方式实现了应用的访问。

任务总结

① 深入理解Kubernetes提供的PV和PVC服务。PV是由管理员创建的，任何用户可以通过创建自己的PVC实现使用PV资源。

② 注意当PVC申请PV时，需要指明PV的容量、访问模式和storageclass名称，名称需要严格按照PV定义的名称。

③ 本任务创建PV方式都是静态创建，也可通过动态方式创建，可参考官网https://kubernetes.io/docs/concepts/storage/storage-classes/。

④ service提供ClusterIP和node:port两种访问方式访问服务，如果采用node:port方式需要在type键中指定NodePort模式。端口号是随机产生的，也可以在service配置文件中的ports键用nodePort指定。

至此，已经对kubernetes入门，更多的kubernetes应用请参考https://kubernetes.io/docs。

任务扩展

本任务创建PV方式都是静态创建，也可通过动态方式创建，读者参考官网https://kubernetes.io/docs/concepts/storage/storage-classes/尝试用动态方式创建。